Environmental Science and Management:
A Statistical Perspective

Environmental Science and Management: A Statistical Perspective

Editor: Bernie Goldman

RCallisto
Reference

www.callistoreference.com

Callisto Reference,
118-35 Queens Blvd., Suite 400,
Forest Hills, NY 11375, USA

Visit us on the World Wide Web at:
www.callistoreference.com

ISBN: 978-1-64116-164-0 (Hardback)

Cataloging-in-Publication Data

Environmental science and management : a statistical perspective / edited by Bernie Goldman.
    p. cm.
Includes bibliographical references and index.
ISBN 978-1-64116-164-0
1. Environmental sciences--Statistical methods. 2. Environmental management--Statistical methods.
3. Environmental protection--Statistical methods. I. Goldman, Bernie.
GE45.S73 E58 2019
363.7--dc23

Table of Contents

Preface

This book was inspired by the evolution of our times; to answer the curiosity of inquisitive minds. Many developments have occurred across the globe in the recent past which has transformed the progress in the field.

Environmental science is an interdisciplinary field that integrates physical, biological and information sciences to study the environment and develop solutions for environmental problems. The study of earth processes, natural resource management, pollution control and mitigation, alternative energy systems and effects of climate change are covered within this domain. The related areas of environmental studies and environmental engineering are significant to the understanding of the human relationship with the environment and for developing strategies for improving environmental quality. The study of the earth's atmosphere, ecology, crust and chemical alterations in the atmosphere are some of the focus areas of environmental science. Management of the influence of human societies on the environment is within the domain of environmental resource management. It takes into consideration all aspects of the biotic and abiotic environment, as well as relationships between living organisms and their habitats. This book contains some path-breaking studies in the field of environmental science and management. It presents researches and studies performed by experts across the globe. It will serve as a valuable source of reference for graduate and postgraduate students.

This book was developed from a mere concept to drafts to chapters and finally compiled together as a complete text to benefit the readers across all nations. To ensure the quality of the content we instilled two significant steps in our procedure. The first was to appoint an editorial team that would verify the data and statistics provided in the book and also select the most appropriate and valuable contributions from the plentiful contributions we received from authors worldwide. The next step was to appoint an expert of the topic as the Editor-in-Chief, who would head the project and finally make the necessary amendments and modifications to make the text reader-friendly. I was then commissioned to examine all the material to present the topics in the most comprehensible and productive format.

I would like to take this opportunity to thank all the contributing authors who were supportive enough to contribute their time and knowledge to this project. I also wish to convey my regards to my family who have been extremely supportive during the entire project.

Editor

A Counting Process with Gumbel Inter-arrival Times for Modeling Climate Data

K.K. Jose,
St.Thomas College, Pala

Bindu Abraham,
B.P.C.II College, Piravom

Abstract

Changes in temperature and rainfall will lead to frequent occurrence of floods, droughts as well as heat and cold waves. In this paper we introduce a new generalized counting process with Gumbel inter-arrival time distribution. This count model can be applied to analyze the climate variability due to the changes in temperature and rainfall. The decreasing hazard function of the new model leads to over dispersion, whereas increasing hazard function leads to underdispersion. Thus this new Gumbel count model can model both over and under dispersed count data. The new model has many nice features such as its closed form nature, computational simplicity, ability to nest Poisson, existence of moments etc. The use of the model is illustrated in two applications with respect to a data on the monthly rainfall of Kerala, the southern state of India which crossed the extreme level during 2005-2008 (under dispersed) and the data on the daily temperature in Kerala which attained the maximum level in Kerala during 2005-2008 (over dispersed).

Keywords: Inter-arrival times, Gumbel distribution, counting process, hazard function, over dispersion, under dispersion.

1. Introduction

The untimely rain in Kerala, which hit the entire region has caused crop damage and flooding. It is estimated that farmers could not harvest paddy worth about Rs.128 crores(1280 million rupees) due to unexpected flooding in the Kuttanad fields extending to 2000 hectares which is quite unusual with the normal summer rain. Experts suggest that this untimely rain is a clear evidence of climate change.

In a report in the news daily, *The Hindu*, it is noted that the "abrupt changes in precipitation and temperature are the main characteristics of climate change. Rising global temperatures will bring changes in weather patterns, rising sea levels and increased frequency and intensity of extreme weather". The summer rain usually comes to the state as a relief to the inhabitants in the month of March and April as it may ease the water shortage. Unlike in temperature trends, rainfall trends are uncertain at several locations see (Kumar et al. 2002, Krishnakumar et al. 2008 and 2009, Rao et al., 2008). It is also a matter of concern that the untimely rain may cause an outbreak of infectious diseases. The state had witnessed infectious diseases in the last few years during rainy season, particularly with Chickungunya and Dengue fever. There had been reports of jaundice attack at many places in the state.(Miguel et al. 2008) studied the influence of temperature and rainfall on the evolution of cholera epidemics in Lusaka, Zambia, 2003-2006. (Pascual et al. 2002) analyzed the quantitative evidence on cholera and climate and (Michael et al. 2008) studied the seasonality of cholera from 1974 to 2005. As this is a clear indication of climate change, both the government and the people must be vigilant to cop up with untimely eventualities which may be a regular phenomenon in the coming years. To study the climate change it is necessary to know the pattern of changes in the level of temperature and rainfall at different time points. Here we take the series of time points which crosses the extreme temperature as well as rainfall . Then by using this time series we developed count models to predict the future climate change.

There is a straight forward connection between the count model and a timing process. But out of many of the count models that have been developed over the years, (see Wimmer and Altman, 1999), very few share this relationship. Poisson count model is truly valid only in the case where the data of interest support the restrictive assumptions of equi-dispersion (i.e., the mean and variance of the data are equal). Statisticians recognized this limitation for many years and now use models that admits over dispersion (i.e., the data sets marked by a fatter, longer right tail than the Poisson can accommodate). A heterogeneous Gamma-Poisson(Negative binomial) model is the first count model invoked for this common situation. Next problem is how to accommodate the data sets with under dispersion. Statisticians have acknowledged and addressed this issue in different ways (see King, 1989; Cameron and Johansson, 1997; Trivedi and Cameron, 1996). But with the possible exception of a count model featuring gamma distributed inter arrival times (see Winklemann, 1995a,b), none of these under dispersed count models offers the conceptual elegance and usefulness of the Poisson exponential connection.

An important entity for the analysis of durations, used to capture duration dependence is the hazard rate h(t) which gives the instantaneous exit probability conditional on survival. Hence we consider, $h(t) = \frac{f(t)}{1-F(t)} = \frac{d}{dt} log F(t)$ where f(t) and F(t) are the density function and cumulative distribution function of the inter arrival time respectively. The hazard function captures the underlying time dependence of the process. A decreasing hazard function implies that the waiting time is less likely to end the longer it lasts. This situation is referred to as negative duration dependence. An increasing hazard function implies that the waiting time is more likely to end the longer it lasts. This situation is referred to as positive duration dependence. No duration dependence corresponds to the case of a con stant hazard. The hazard is a constant if and only if the distribution of waiting times is exponential.

We assume that the waiting times between the events are independent but not exponential (which would lead to the Poisson distribution for counts). Instead they follow some other distribution with a nonconstant hazard function. If the hazard function is a decreasing function of time, the distribution displays negative duration dependence. If the hazard function is an increasing function of time, the distribution displays positive duration dependence. In both cases, the conditional probability of a current occurrence depends on the time since the last occurrence rather than on the number of previous events. Events are dependent in the sense that the occurrence of at least one event (in contrast to none) up to time t influences the probability of a further occurrence in $t + \Delta t$.

There is a link between duration dependence and dispersion. If we denote the mean of the interarrival distribution by μ, the variance by $\sigma,^2$ then we say that the distribution has negative duration dependence if $\frac{d}{dt}h(t) < 0$ and positive duration dependence if $\frac{d}{dt}h(t) > 0$. If the hazard function is monotonic, then we have if $\frac{d}{dt}h(t) > 0$, then $\frac{\sigma}{\mu} < 1$; if $\frac{d}{dt}h(t) = 0$, then $\frac{\sigma}{\mu} = 1$; if $\frac{d}{dt}h(t) < 0$, then $\frac{\sigma}{\mu} > 1$, (see Bradlow et. al, 2002). These three cases correspond to count data characterized by under dispersion, equi dispersion and over dispersion respectively. It is shown that negative duration dependence (asymptotically) causes over-dispersion and positive duration dependence causes under-dispersion. The Poisson process can be taken as a sequence of independently and identically exponentially distributed waiting times (see Cox, 1972) . To derive a generalized model we replace the exponential distribution with a less restrictive non negative distribution. Possible candidates known from the duration literature are the Weibull (see McShane et.al, 2008), the gamma(including generalized gamma) (see Winkelmannn, 1995a,b, 2008), and the log normal distributions (see Bradlow et. al, 2002; Everson and Bradlow, 2002; Miller et al., 2006). Both Weibull and gamma nest the exponential distribution and both allow for a monotone hazard rate function that is duration dependent. Recently some authors developed (see Jose and Bindu, 2011) a count model with Mittag-Leffler inter-arrival time distribution.

In this paper we replace the exponential distribution by Type II Gumbel distribution where Weibull and Frechet are the special cases. A corresponding count model is formulated which nest Poisson process as a special case. The Gumbel inter arrival time model is richer than the exponential, because it allows for nonconstant hazard rates (duration dependence). We can derive the model by using a polynomial expansion.The remainder of this article is as follows. In section 2, a description about Type II Gumbel distribution is given. Section 3 contains the derivation of the Type II Gumbel count model, focusing on the polynomial expansion that leads to the closed form benefits and its prperties. In section 4 the Gumbel count model is used in two real applications in climate data.

2. Gumbel Distribution

The Gumbel distribution was first developed by a German mathematician Emil Gumbel (1891-1966). Gumbel's focus was primarily on applications of extreme value theory to engineering problems, in particular modeling of meteorological phenomena such as annual flood flows. The Gumbel distribution, also known as the Extreme Value Type I distribution, is unbounded (defined on the entire real axis). The cumulative distribution function is

$$F(x; a, b) = 1 - e^{-bx^{-a}}.$$

The moments $E[X^k]$ exist for $k < a$. We can define Type II Gumbel distribution, as one whose cumulative distribution function [c.d.f] has the polynomial expansion of the form

$$F(x; a, b) = \sum_{j=1}^{\infty} \frac{(-1)^{j+1}(bx^{-a})^j}{\Gamma(j+1)} \tag{1}$$

and p.d.f is given by

$$f(x; a, b) = \sum_{j=1}^{\infty} \frac{(-1)^{j+1}(-a)jb^j(x^{-aj-1})}{\Gamma(j+1)}. \tag{2}$$

The type II Gumbel distribution admits a closed form expression for the hazard function as

$$h(t) = \frac{f(t)}{1 - F(t)} = -abx^{-a-1}.$$

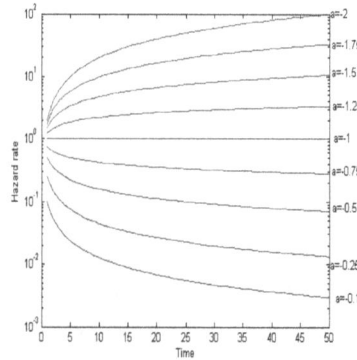

Figure 1

Hazard rate of Type II Gumbel distribution for different values of a and b=0.7.

which is monotonically increasing for $a < -1$, monotonically decreasing for $a > -1$, and constant when $a = -1$. Figure 1 supports this intuitive fact.

Gumbel has shown that the maximum value (or first order statistic) in a sample of a random variable following an exponential distribution approaches closer and closer to the Gumbel distribution with increasing sample size. In hydrology, therefore, the Gumbel distribution is used to analyze such variables as monthly and annual maximum values of daily rainfall and river discharge volumes.

3. Gumbel count model

Let Y_n be the time from the measurement origin at which the nth event occurs. Let $X(t)$ denote the number of events that have occurred upto time t. The relationship between interarrival times and the number of events is

$$Y_n \leq t \quad \Leftrightarrow \quad X(t) \geq n.$$

We can derive the type II Gumbel count model by using the following relationship

$$\begin{aligned} G_n(t) = P[X(t) = n] &= P[X(t) \geq n] - P[X(t) \geq n+1] \\ &= P[Y_n \leq t] - P[Y_{n+1} \leq t]. \end{aligned} \tag{3}$$

If we let the cumulative density function(cdf) of Y_n be $F_n(t)$ then,

$$G_n(t) = P[X(t) = n] = F_n(t) - F_{n+1}(t). \tag{4}$$

In the case where the measurement time origin (and thus counting process) coincides with the occurrence of an event, $F_n(t)$ is simply the n-fold convolution of the common interarrival time distribution which may or may not have a closed form solution. Based upon (3) we can derive our Type 2 Gumbel count model using the polynomial expansion of F(t).

$$G_n(t) = \int_0^t F_{n-1}(t-s)f(s)ds - \int_0^t F_n(t-s)f(s)ds = \int_0^t G_{n-1}(t-s)f(s)ds. \qquad (5)$$

Before proceeding to develop the general solution to the problem, we note that $F_0(t)$ is 1 and $F_1(t) = F(t)$ for every t. Therefore we have

$$G_0(t) = F_0(t) - F_1(t) = 1 - F(t) = e^{-bx^{-a}} = \sum_{j=0}^{\infty} \left[\frac{(-1)^j (bx^{-a})^j}{\Gamma(j+1)} \right]. \qquad (6)$$

By equation (5) we can derive

$$G_1(t) = \int_0^t G_0(t-s)f(s)ds$$

$$= \int_0^t \left(\sum_{j=0}^{\infty} \frac{(-1)^j (b(t-s)^{-a})^j}{\Gamma(j+1)} \right) \times \left(\sum_{k=1}^{\infty} \frac{(-1)^{k+1} - akb^k s^{-ak-1}}{\Gamma(k+1)} \right) ds$$

$$= \sum_{j=0}^{\infty} \sum_{k=1}^{\infty} \frac{(-1)^j (-1)^{k+1} b^j b^k}{\Gamma(j+1)\Gamma(k+1)} \int_0^t -ak(t-s)^{-aj} s^{-ak-1} ds$$

$$= \sum_{j=0}^{\infty} \sum_{k=1}^{\infty} \frac{(-1)^j (-1)^{k+1} b^j b^k}{\Gamma(j+1)\Gamma(k+1)} \times \frac{t^{-ak} t^{-aj} \Gamma(-aj+1)\Gamma(-ak+1)}{\Gamma(-ak+-aj+1)}.$$

Then by using a change of variables m=j and l=m+k we obtain

$$G_1(t) = \sum_{l=1}^{\infty} \left(\sum_{m=0}^{l-1} \frac{(-1)^m (-1)^{l-m+1} b^m b^{l-m}}{\Gamma(m+1)\Gamma(l-m+1)} \times \frac{t^{-a(l-m)} t^{-am} \Gamma(-am+1)\Gamma(-a(l-m)+1)}{\Gamma(-a(l-m)-am+1)} \right)$$

$$= \sum_{l=1}^{\infty} \left(\sum_{m=0}^{l-1} \frac{(-1)^{l+1} (bt^{-a})^l \Gamma(-am+1)\Gamma(-al+am+1)}{\Gamma(m+1)\Gamma(l-m+1)\Gamma(-al+1)} \right)$$

$$= \sum_{l=1}^{\infty} \frac{(-1)^{l+1} (bt^{-a})}{\Gamma(-al+1)} \left(\sum_{m=0}^{l-1} \frac{\Gamma(-am+1)\Gamma(-al+am+1)}{\Gamma(m+1)\Gamma(l-m+1)} \right)$$

$$= \sum_{l=1}^{\infty} \frac{(-1)^{l+1} (bt^{-a}) \delta_m^l}{\Gamma(-al+1)}, \text{ where } \delta_m^l = \sum_{m=0}^{l-1} \frac{\Gamma(-am+1)\Gamma(-al+am+1)}{\Gamma(m+1)\Gamma(l-m+1)}.$$

$$\text{Similarly } G_2(t) = \sum_{l=2}^{\infty} \frac{(-1)^{l+2} (bt^{-a}) \delta_l^2}{\Gamma(-al+1)}.$$

We use the method of mathematical induction to derive $G_n(t)$. Thus we have

$$G_n(t) = \sum_{l=n}^{\infty} \frac{(-1)^{l+n}(bt^{-a})\delta_l^n}{\Gamma(-al+1)}.$$

$$G_{n+1}(t) = \int_0^t G_n(t-s)f(s)ds$$

$$= \int_0^t \left(\sum_{j=n}^{\infty} \frac{(-1)^{j+n}(bt^{-a})\delta_j^n}{\Gamma(-aj+1)} \right) \times \left(\sum_{k=1}^{\infty} \frac{(-1)^{k+1} - akb^k s^{-ak-1}}{\Gamma(k+1)} \right) ds$$

$$= \sum_{j=n}^{\infty} \sum_{k=1}^{\infty} \frac{(-1)^{j+n}(-1)^{k+1}b^j b^k \delta_j^n}{\Gamma(-aj+1)\Gamma(k+1)} \times \frac{t^{-ak}t^{-aj}\Gamma(-aj+1)\Gamma(-ak+1)}{\Gamma(-ak+-aj+1)}$$

$$= \sum_{l=n+1}^{\infty} \frac{(-1)^{l+n+1}(bt^{-a})^l}{\Gamma(-al+1)} \left(\sum_{m=n}^{l-1} \delta_m^n \frac{\Gamma(-al+am+1)}{\Gamma(l-m+1)} \right)$$

$$= \sum_{l=n+1}^{\infty} \frac{(-1)^{l+n}(bt^{-a})^l \delta_l^{n+1}}{\Gamma(-al+1)}, \text{ where } \delta_l^{n+1} = \sum_{m=n}^{l-1} \delta_m^n \frac{\Gamma(-al+am+1)}{\Gamma(l-m+1)}.$$

Theorem 3.1 *If the interarrival times are independently and identically distributed as Type II Gumbel distribution, then the count model probabilities are given by*

$$G_n(t) = P[X(t) = n] = \sum_{j=n}^{\infty} \frac{(-1)^{j+n}(bt^{-a})^j \delta_j^n}{\Gamma(-aj+1)}, a < 0, n = 0, 1, 2... \tag{7}$$

$$where \ \delta_j^0 = \frac{\Gamma(-aj+1)}{\Gamma(j+1)}, j = 0, 1, 2, ...$$

$$and \ \delta_j^{n+1} = \sum_{m=n}^{j-1} \delta_m^n \frac{\Gamma(-aj+am+1)}{\Gamma(j-m+1)}, for \ n = 0, 1, 2...for \ j = n+1, n+2, n+3, ...$$

3.1. Characteristics of the Gumbel count model

1. The model handles both over-dispersed and under-dispersed data

   Through extensive simulations we have verified that the hazard function of the Type II Gumbel distribution is a decreasing function of time when $a > -1$, so that the distribution displays negative duration dependence which causes over-dispersion, whereas $a < -1$ the hazard function is monotonically increasing function of time, so that the distribution displays positive duration dependence which causes under-dispersion. A lack of duration dependence occurs when $a = -1$ which leads to the Poisson distribution with equal mean and variance.

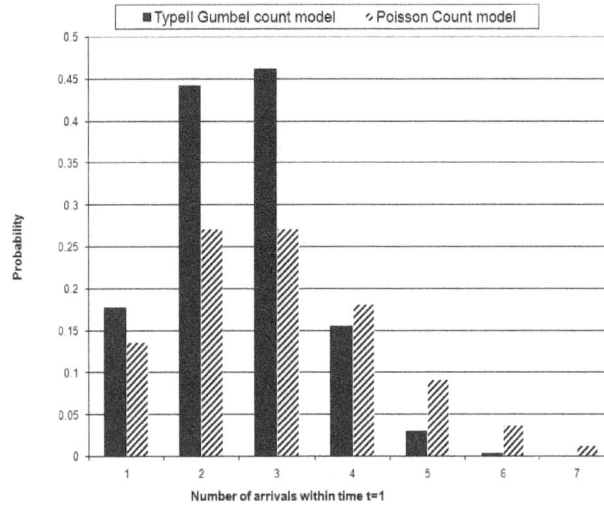

Figure 2

Probability histogram of Type II Gumbel count model ($a = -1.5$) and Poisson model ($\lambda=2$) displaying under-dispersion.

Figures 2 and 3 display probability histograms for the Type II Gumbel and Poisson models with different parameter values. Both the Gumbel and Poisson have identical means, but their dispersion is quite different. In Figure 2, probability histogram for an under-dispersed Type II Gumbel count model with parameters $a = -1.5$ and $b = 1.845$ and a Poisson with $\lambda = 2$. The variance of the Gumbel count model in this case is 0.388 which is smaller than the mean, as expected. In Figure 3, we have the probability histogram for an over-dispersed Type II Gumbel count model with parameters $a = -0.5$ and $b = 0.89$, and again follow the Poisson with $\lambda = 2$. The variance of the Gumbel count model in this case is 2.17 which is greater than the mean, as expected.

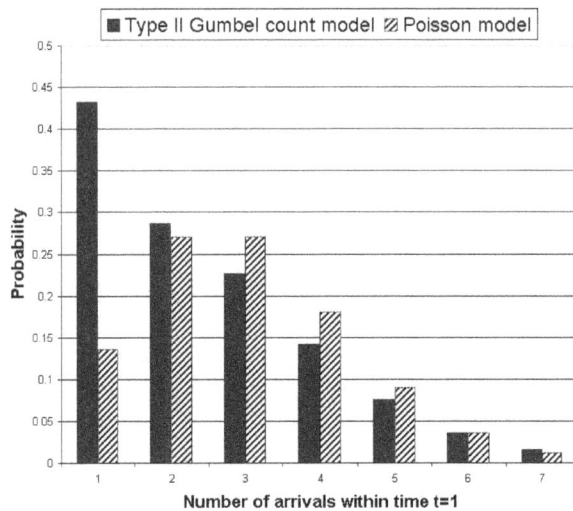

Figure 3

Probability histogram of Type II Gumbel count model ($a = -0.5$) and Poisson model ($\lambda=2$) displaying over-dispersion.

2. The model generalizes the most commonly used model such as the Poisson as special case when $t = 1$ and $a = -1$. Then

$$G_n(t) = P(X(t) = n) = \sum_{j=n}^{\infty} \frac{(-1)^{j+n} b^j \delta_j^n}{\Gamma(j+1)}.$$

This is a Poisson process with unit rate.

3. The mean and variance of the Type II Gumbel count model exist, when $a < 0$

$$\text{Mean} = E[X(t)] = \sum_{n=1}^{\infty} \sum_{j=1}^{\infty} n \frac{(-1)^{j+n} (bt^{-a})^j \delta_j^n}{\Gamma(-aj+1)}.$$

$$\text{Variance} = Var[X(t)] = \sum_{n=1}^{\infty} \sum_{j=1}^{\infty} n^2 \frac{(-1)^{j+n} (bt^{-a})^j \delta_j^n}{\Gamma(-aj+1)} - \left(\sum_{n=1}^{\infty} \sum_{j=1}^{\infty} n \frac{(-1)^{j+n} (bt^{-a})^j \delta_j^n}{\Gamma(-aj+1)} \right)^2.$$

Table 1
Values of probabilities, mean and variance of the Type II Gumbel count model for different values of the parameter a when b=0.7 at t=1

a	$G_0(t)$	$G_1(t)$	$G_2(t)$	$G_3(t)$	$G_4(t)$	$G_5(t)$	$\cdots$	μ	σ^2	relation
-2	0.6999	0.4722	0.0700	0.0035	0.0004	0.0001	$\cdots$	0.6230	0.3969	under dispersion
-1.75	0.6350	0.4531	0.0899	0.0069	0.0003	0.0000	$\cdots$	0.6548	0.4509	under dispersion
-1.5	0.5728	0.4262	0.1127	0.0132	0.0009	0.0000	$\cdots$	0.6950	0.5285	under dispersion
-1.25	0.5218	0.3913	0.1364	0.0240	0.0028	0.0002	$\cdots$	0.7487	0.6439	under dispersion
-1	0.4903	0.3516	0.1578	0.0413	0.0077	0.0011	$\cdots$	0.8282	0.8246	equi dispersion
-0.75	0.4826	0.3136	0.1733	0.0651	0.0191	0.0047	$\cdots$	0.9614	1.1273	overdispersion
-0.5	0.4966	0.2857	0.1821	0.0931	0.0409	0.0160	$\cdots$	1.2074	1.6556	overdispersion
-0.25	0.5259	0.2745	0.1884	0.1202	0.0725	0.0417	$\cdots$	1.6490	2.4248	overdispersion
-0.1	0.5483	0.2776	0.1938	0.1334	0.0906	0.0609	$\cdots$	1.9758	2.7823	overdispersion

Table 1 gives the probabilities of Type II Gumbel count model for different values of the parameter a when $b = 0.7$. By simulation of the Type II Gumbel count model we verified that for $a > -1$, the conditional variance exceeds conditional expectation which represents the over-dispersion and for $a < -1$, the conditional expectation exceeds conditional variance which represents the under-dispersion but for $a = -1$ conditional mean equals conditional variance which means the equi-dispersion. Thus Type II Gumbel count model can be used to represent overdispersed, under-dispersion as well as equidispersed real data. Table 1 supports this intuitive fact.

4. If the inter arrival times of the data set are Type II Gumbel distributed then, we have a corresponding counting model to use. The model (7) is derived from Type II Gumbel timing model, the link between the timing model and its counting model equivalent is maintained. Hence in those cases where an analysis of the inter-arrival times suggests that a more flexible timing model is needed, it can now be incorporated via its count model equivalent. Furthermore, in those cases where one only has count data, but would like to make forecasts of the next arrival time, this can be done given the timing and count model link that is now achieved.

5. We can simulate the Type II Gumbel count model. The model is computationally feasible to work with and it is estimable without requiring a formal programming language or time consuming simulation based methods.

4.   Application to the real data sets

Although Kerala lies close to the equator, its proximity with the sea and the presence of the fort like Western Ghats, provides it with an equable climate which varies little from season to season. The temperature varies from 28^0 to 32^0C. Southwest Monsoon and Retreating Monsoon (Northeast Monsoon) are the main rainy seasons. The thermo-sensitive crops like black pepper, cardamom, tea, coffee and cocoa will be badly accepted as temperature range (the difference between maximum and minimum temperatures) is likely to increase and rainfall is likely to decline.

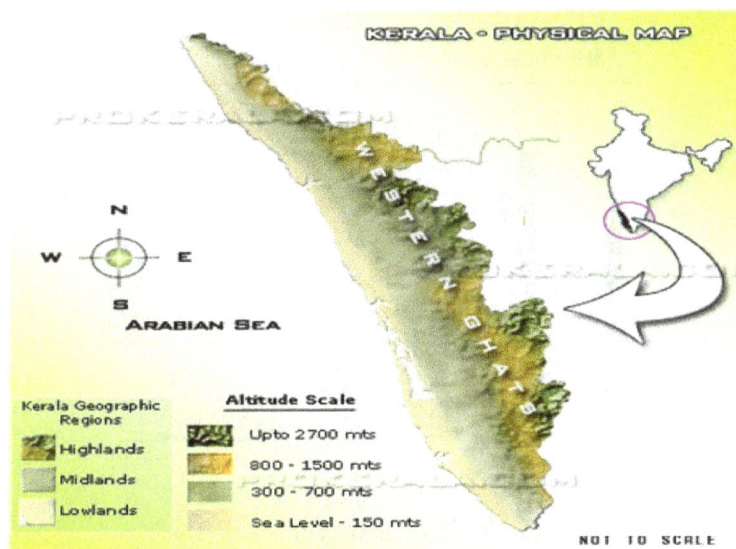

Figure 4
The location of Kerala

4.1. Monthly Rainfall Data in Kerala(Under dispersed case)

In this section we apply the model to a data on the monthly rainfall of Kerala which crosses the extreme level. The data collected concern monthly distribution of Normal and Actual rainfall of Kerala state in India from 1994-2003. The total annual rainfall in the state varies from 360cm over the extreme northern parts to about 180cm in the southern parts. The southwest monsoon (June to October) is the principal rainy season in the state. The thunderstorm rains in the pre monsoon months of April and May and that of monsoon months are locally known as 'Edavapathi'. Rainfall during northeast monsoon season is known as 'Thulavarsham' in local language. The southwest monsoon sets over the southern parts of the State by about 1st June and extends over the entire State by 5th June. June and July are the rainiest months, each accounting individually to about 23 of annual rainfall. In each of these months number of rainy days (with daily rainfall of at least 2.5 mm) varies from 27 in the north to 15 in the south.

The time interval between the months having extreme rainfall had the conditional mean 1.5217 and conditional variance 1.3662. Thus the data set is underdispersed and hence we can apply the Gumbel count model with estimates of parameters $\hat{a} = -1$ and $\hat{b} = 0.4450$.

To test whether there is significant difference between an observed interarrival time distribution and the Gumbel distribution we use Kolmogorov-Smirnov [K.S] test for H_0 : Gumbel distribution with parameter $a = -1$ and $b = 0.4450$ is a good fit for the given data. Here the calculated value of the K.S. test statistic is 0.0869 and the critical value corresponding to the significance level 0.01 is 0.2403 showing that the Gumbel assumption for inter arrival times is valid.

To estimate the number of months having extreme rainfall in a class, we use the Gumbel Count model. The Figure 5 shows that the Gumbel count model can be applied to underdispersed data.

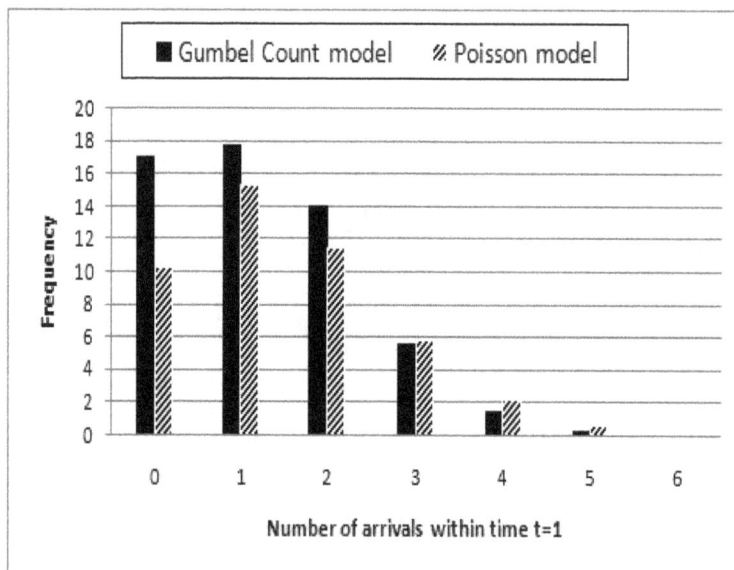

Figure 5

Probability histogram of the expected number of months having extreme rainfall in Kerala during 1994-2003 according to Gumbel count model and Poisson model

4.2. Data on Maximum Temperature in Kerala (Over dispersed data)

In this section we apply the model to a data on the daily maximum temperature which attains extreme during 2005-2008 in Kerala. The data collected is from the meteorological observations recorded at Kayamkulam station during the period 2005-2008.

Day temperatures are more or less uniform over the plains throughout the year except during monsoon months when these temperatures drop down by about 3 to 5^0C. Both day and night temperatures are lower over the plateau and at high level stations than over the plain. Day temperatures of coastal places are less than those of interior places. March is the hottest month with a mean maximum temperature of about 33^0C. Mean maximum temperature is minimum in the month of July when the State receives plenty of rainfall and the sky is heavily clouded. It is 28.5^0C for the State as a whole in July, varying from about 28^0C in the north to about 29^0C in the South. The night temperature is minimum in January when clouding is also minimum. For the State as a whole it is about 22.5^0 C in January, varying from 22^0C in the north to 22.6^0C in the South). The time interval between (interarrival) the days having maximum temperature had the conditional mean 1.7553 and the conditional variance 6.2178.

Thus the data set is overdispersed, and hence we can apply the Gumbel Count model with estimates of parameters $\hat{a} = -0.93$ and $\hat{b} = 0.5950$.

To test whether there is significant difference between an observed interarrival time distribution and the Gumbel distribution, we use Kolmogorov-Smirnov [K.S] test for H_0 : Gumbel distribution with parameters $a = -0.93$ and $b = 0.5950$ is a good fit for the given data. Here the calculated value of the K.S. test statistic is 0.1711 and the critical value corresponding to the significance level 0.01 is 0.1869 showing that the Gumbel assumption for inter arrival times is valid.

To estimate the number of days having maximum temperature in a class, we use the Gumbel Count model. The Figure 6 shows that the Gumbel count model can be applied to overdispersed data.

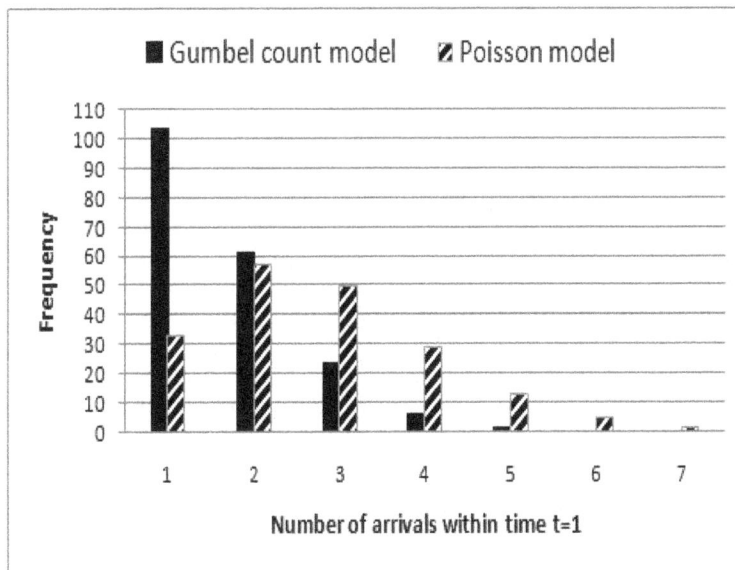

Figure 6

Probability histogram of the expected number of days having maximum temperature in Kerala during 2005-2008 according to Gumbel count model and Poisson model

5. Conclusions

In this article we have introduced a new count model based upon Gumbel inter arrival time process. More importantly, the model provided a sizeable improvement over the more traditional Poisson. The new model can be used to predict the extreme changes in temperature and rainfall which causes the climate change. One important advantage of the new model is that it removed the artificial symmetry between overdispersion and equidispersion, a violation of the constant hazard assumption underlying the Poisson model. This new model can be treated as a generalization of the Poisson distribution. The new model has closed form nature and computation is possible using Matlab. This new model can be applied to real data sets where the assumption of equidispersion is violated.

Acknowledgements

The first author thanks the International Statistical Institute, Netherlands for providing travel support under World Bank Fund Scheme to attend and present this paper in the TIES 2012 International Conference held at Hyderabad, India. The authors are also grateful to Bovas Abraham, University of Waterloo, Canada and the editors for their valuable comments and suggestions.

References

Bradlow ET, Hardie BGS, Fader PS (2002). "Bayesian Inference for the Negative Binomial Distribution via Polynomial Expansions". *Journal of Computational and Graphical Statistics*, **11**, 189-201.

Cameron AC, Johansson P (1997). "Count Data Regression Using Series Expansion: With Applications". *Journal of Applied Econometrics*, **12**, 203-223.

Cox DR (1972). "Regression Models and Life Tables". *Journal of the Royal Statistical Society*, Ser.B, **34**,187-220.

Everson PJ, Bradlow ET (2002). "Bayesian Inference for the Beta binomial distribution via polynomial expansion". *Journal of Computational and Graphical Statistics*, **11**, 202-207.

Gumbel EJ (1954). "Statistical theory of extreme values and some practical applications". *Applied Mathematics Series*, **33**.

Jose KK, Bindu A (2011). "A count model based on Mittag-Leffler inter arrival times". *Statistica*, anno LXXI, **4**, 501-514.

King G (1989). "Variance Specifications in Event Count Models: From Restrictive Assumptions to a Generalized Estimator". *American Journal of Political Science*, **33**, 762-784.

Krishnakumar KN, Rao HS, Gopakumar CS (2008). "Climate change at selected locations in the humid tropics". *Journal of Agrometeorology.*, **10**(1), 59-64.

Krishnakumar KN, Rao HS, Gopakumar CS (2009). "Rainfall trends in twentieth century over Kerala, India". *Atmospheric Environment* **43** 1940-1944.

Kumar KR, Kumar KK, Ashrit RG, Patwardhan SK, Pant GB (2002). "Climate change in India: Observation and Model Projections". *Climate change and India-Issues, Concern and Opportunities. (eds. P.R. Shukla, K.S. Sharma, and P.V. Ramana). Tata McGraw-Hill Publishing Company Limited, New Delhi,* pp 24-75.

Mcshane B, Adrian M, Bradlow ET, Fader PS (2008). "Count models based on Weibull interarrival times". *Journal of Business and Economic Statistics*, **26**(3), 369-378.

Michael E, Caryl F, Islam MS, Mohammad A (2008). "Seasonality of cholera from 1974 to 2005: a review of global patterns". *International Journal of Health Geographics*, **7**(31).

Miguel LF, Bauernfeind A, Jimenez JD, Gil CL, Omeiri NE, Guibert DH (2008). "Influence of temperature and rainfall on the evolution of cholera epidemics in Lusaka, Zambia, 2003-2006: analysis of a time series". *Transactions of Royal Society of Tropical Medicine and Hygiene*, doi:10.1016/j.trstmh.2008.07.017.

Miller SJ, Bradlow ET,Dayaratna K (2006). "Closed form Bayesian inferences for the logit model via polynomial expansions". *Quantitative Marketing and Economics*, **4**, 173-206.

Pascual M, Menno JB, Andrew PD (2002). "Cholera and climate: revisiting the quantitative evidence". *Microbes and Infection*, **4**, 237-245.

Rao HS, Ram M, Gopakumar CS, Krishnakumar KN (2008). "Climate Change and cropping systems over Kerala in the humid tropics". *Journal of Agrometeorology, special issue* Part 2: 286-291.

Trivedi PK, Cameron AC (1996). *Applications of Count Data Models to Financial Data.* Handbook of Statistics, North Holland, Chap. **12**, pp. 363-391.

Wimmer G, Altmann G (1999). *Thesaurus of Univariate Discrete Probability Distributions*, Germany, Stamm Verlag.

Winkelmann R (1995a). "Duration dependence and dispersion in count data models". *Journal of Business and Economic Statistics*, **13**, 467-474.

Winkelmann R (1995b). "Recent developments in count data modeling theory and applications". *Journal of Economic Survey*, **9**, 1-24.

Winkelmann R (2008). *Econometric Analysis of Count Data*, Springer 5th Edition.

Affiliation:

K. K. Jose,
Professor, Department of Statistics, St.Thomas College, Pala, Mahatma Gandhi University, Kottayam, Kerala-686574 India.
Email: (kkjstc@gmail.com)

Bindu Abraham,
Assistant Professor, Department of Statistics, Baselios Poulose II Catholicose College, Piravom, Mahatma Gandhi University, Kottayam, Kerala-686664 India.
Email: (babpc@rediffmail.com)

2

Modeling the Impact of Afforestation on Global Climate: A 2-Box EBM

Craig Jackson
Department of Mathematics
and Computer Science,
Ohio Wesleyan University

Sriharsha Masabathula
Department of Economics,
Ohio Wesleyan University

Abstract

Afforestation programs have become increasingly prevalent around the world as trees are considered crucial in mitigating climate change due to their carbon sequestration potential. In recent years, international agreements such as the Clean Development Mechanism established under the United Nations Framework Convention for Climate Change have notably fueled afforestation activities. However, several complicating factors are often neglected when evaluating the effects of afforestation on global climate. For instance, while carbon uptake by forests reduces the greenhouse effect, the increase in evapotranspiration due to afforestation tends to increase it. An increase in forest cover also lowers the albedo of afforested regions due to the fact that afforestation efforts tend to be carried out on barren lands having relatively high albedo. Further, atmospheric transport exacerbates the cumulative effect of afforestation on global temperatures due to the interaction of poleward transport of sensible and latent heat with ice-albedo feedback.

In this study, we assess the impact of afforestation on global and regional temperatures utilizing a mathematical climate model incorporating carbon dioxide forcing, land/ice albedo feedback, evapotranspiration, and atmospheric heat transport. We investigate the extent to which changes in surface reflectivity and moisture content of the atmosphere caused by afforestation offset the cooling potential of carbon sequestration. In addition, we examine the degree to which these climatic responses depend on the latitude of the afforested region. Considerations such as these have the potential to increase the positive impact of afforestation efforts by identifying land types and latitude regions that, when planted, result in greater mitigation of global warming.

Keywords: afforestation, conceptual climate models.

1. Introduction

The United Nations Conference on Environment and Development, held in 1992, has been successful in generating widespread awareness regarding the need for sustainable development. Following the conference, international market-based development mechanisms, such as the Clean Development Mechanism, were established under the United Nations Framework Convention on Climate Change and have notably fueled afforestation activities. In addition, initiatives such as the Billion Tree Campaign, led by the United Nations Environment Programme, have raised awareness about the carbon-sequestration potential of large-scale plantations, leading to increased interest in afforestation as an effective means of climate change mitigation.

However, a growing body of literature has consistently challenged advocates of afforestation-driven carbon sequestration with the admonition that afforestation can potentially result in a net positive radiative forcing resulting in an overall warming of the global climate. Gibbard et al. (2005) observe that "when changing from grass and croplands to forest, there are two competing effects of land cover change on climate: an albedo effect which leads to warming and an evapotranspiration effect which tends to produce cooling" (p. 1). In this same context, Bonan et al. (2008) argues that while boreal forests create a positive forcing due to a low albedo, tropical forests create a negative forcing through evaporative cooling.

It should be noted, however, that the cooling due to evaporation is local and may only be confined to the region of plantation. In general, the moisture added to the atmosphere from increased surface evaporation will be transported to higher latitudes where it condenses, thereby releasing heat. A number of studies have shown water vapor and atmospheric transport of latent heat are very important in shaping the polar amplified response of the climate to forcing (Flannery 1984, Schneider et al. 1997; Alexeev 2003; Rodgers et al. 2003; Alexeev et al. 2005; Langen and Alexeev 2005; Cai 2005; Langen and Alexeev 2007; Graversen and Wang 2009). Therefore, an increase in poleward latent heat transport due to large scale afforestation in the tropics has the potential to lead to extra-tropical warming despite the cooling effect caused by carbon uptake in the forest. In fact, interactions between the atmospheric transport and the ice/snow albedo feedback could lead to an overall global warming response to afforestation.

Because of its strong greenhouse effect and positive dependence on atmospheric temperature the presence of water vapor in the climate system results in a strong positive feedback independent of transport (Hall and Manabe 1999; Held and Soden 2000). As such, the impact on atmospheric water vapor should be taken into account when assessing the potential climate response to large scale afforestation, not simply the negative forcing due to carbon uptake. For instance, Soden et al. (2002) showed that atmospheric drying resulting from a simulated volcanic eruption in a GCM amplifies the negative radiative forcing due to injected volcanic aerosols. Since large afforestation projects on dry/barren lands lead to increased evapotranspiration, there is the possibility that the water vapor feedback in this case will dampen, or even negate, the radiative forcing (Pielke et al. 2002).

Changes in the albedo of afforested regions should also be taken into account given that many large afforestation projects are carried out on lands that have a relatively high albedo. This decrease in albedo due to afforestation will have a local warming effect, though, again, the potential exists for this local effect to be felt in different latitudes due to the atmospheric circulation. Cess (1978) has shown that extremely long term changes in the surface albedo can double the sensitivity of the global climate to factors which produce climate change.

Betts (2000) simulated radiative forcings due to changes in land surface albedo and argues that for boreal forests the positive forcing induced by the decreases in albedo can fully offset the negative forcing induced by carbon sequestration so that afforestation in high latitudes can lead to warming. In such cases, it may even be argued that deforestation is a preferred strategy for mitigating climate change. Foley et al. (2005), relying on Bonan et al. (1992), says that deforestation in high latitudes can cool the climate due to an increase in surface albedo.

Both high-altitude and high-latitude regions – where one can expect consistent snow cover – have been accepted as regions where deliberate land-use change in the form of afforestation can lead to a net positive forcing due to a lower surface albedo. In addition, GCM simulations by Gibbard et al. (2005) showed that total replacement of current vegetation by trees would lead to warming similar to $2 \times CO_2$ scenarios while replacement of vegetation by grassland would lead to moderate cooling. Their simulations also indicate that mid-latitude forestation shows the possibility of a potential positive forcing and net warming.

It is evident, then, that there is more than one effect of land-use change on the global climate. In essence, afforestation can do more than simply sequester carbon. Further, the relative effects of the different feedback processes involved are not well understood and are difficult to estimate using GCMs. Even very large plantations are negligible when compared to the land area that is currently forested. Hence, extracting a simulated climate response to a given plantation in a GCM will be problematic given the inherent variability on multiple timescales that exists in most large models.

Forests have complex non-linear interactions with the atmosphere and affect planetary energetics, the hydrological cycle, and atmospheric composition which can dampen or amplify anthropogenic climate change. An additional complication that is important for modeling, as well as model validation, relates to carbon stock assessment in the field. Unfortunately, different assessment technologies sometimes give different estimates of carbon content. Hence, a consensus view on the best methods to use to gauge carbon stocks has not materialized. Of course, it is even more difficult, if not impossible, to develop accurate general formulae for biomass carbon densities across the board (Christie and Scholes, 1995). This seems to call for a more regional, project-specific approach while evaluating afforestation activities.

Most of our current understanding about forests and their interactions with the climate system comes from models, which are abstractions of many complex systems in our atmosphere. It is these models that contribute to policy making under treaties such as the Kyoto Protocol. Because of this, accurate quantification/parametrization of model processes is essential if the policies we enact are to have the effect we intend them to have. But we would add that a knowledge of important climate processes, including the mechanics of their interaction, is of equal importance in shaping policy going forward.

As models become more complex they allow for greater climate prediction, but they also become less useful for understanding and conceptualizing climate systems. For this reason, we chose to consider the impact of afforestation on the global climate using a simplified 2-box energy balance model. The model is sufficiently detailed in that it incorporates the main climate processes governing the interaction between forests and the climate system as discussed above. However, the model's simplicity (it has only two prognostic variables) allows for a focused study of competing climate feedbacks via a qualitative analysis. Also, individual climate processes can be easily switched off in the model to isolate their effect by means of, for instance, a formal feedback analysis.

Afforestation is widely recognized for its carbon sequestration potential in the policy-world. We think this recognition should be expanded to include albedo changes, water vapor feedbacks, and atmospheric transport of heat. We echo the contention of Pielke et al. (2002) that a system which takes regional effects into account in a new metric will be useful in developing a more comprehensive protocol than what we have currently. A more complete assessment and understanding of the ways in which afforestation can impact the climate system can only aid in our ability to craft sound policies for guiding the implementation of large scale afforestation efforts so they have their intended effect.

2. Model Schematics

The model used in this paper is adapted from a 2-box energy balance model used by Alexeev and Jackson (2013) to assess the relative roles of atmospheric heat transport (AHT) and surface albedo feedback (SAF) in shaping the polar amplified response of the global climate to uniform forcing. It consists of two boxes or regions, shown schematically in Figure 1, one topical and one extra-tropical, dividing the hemisphere equally area-wise at 30°N. Each box contains equal parts land and ocean. The model incorporates surface albedo feedback, atmospheric heat transport, CO_2-dependent emissivity, evapotranspiration, and water vapor feedback in the simplest possible formulation. The change in temperature of the regions is modeled as a function of incoming shortwave solar fluxes, atmospheric heat fluxes (sensible and latent), outgoing longwave radiation, and CO_2 forcing. The model takes into account albedo of the regions as well. Moisture availability for latent heat transport depends on the temperature of the tropical atmosphere as well as the area available for evapotranspiration. We assume free evapotranspiration over both ocean and forested regions, while barren (non-forested) lands are assumed to be dry.

3. Model Equations

Model state variables are T_1 and T_2, the average temperature of the tropical and extra-tropical boxes, respectively. The temperature of the tropical box is assumed to be independent of latitude, while the temperature of the extra-tropical box is assumed to decreases linearly from T_1 at 30°N. This assumption is justified by the annual zonally averaged meridional temperature profile described, say, in Piexoto and Oort (1992). The extent of the ice cap is determined as the area north of the latitude where the temperature crosses a prescribed freezing temperature, here taken to be −2°C. Model equations are given by an energy balance:

$$H \, dT_1/dt = S_1 - F - (A + B \, T_1) + \varepsilon$$
$$H \, dT_2/dt = S_2 \, (1 - 2\alpha a) + F - (A + B \, T_2) + \varepsilon \tag{1}$$

Here H is the heat content of each region, determined primarily by the upper ocean layer heat content; S_1 and S_2 are the net incoming solar fluxes in the tropical and extra-tropical boxes, respectively; A and B are the Budyko-Sellers constants for parametrization of the outgoing long-wave radiation as a function of surface temperature; a is the fractional area of the hemisphere covered by snow/ice; α is the effective ice albedo; and ε is represents

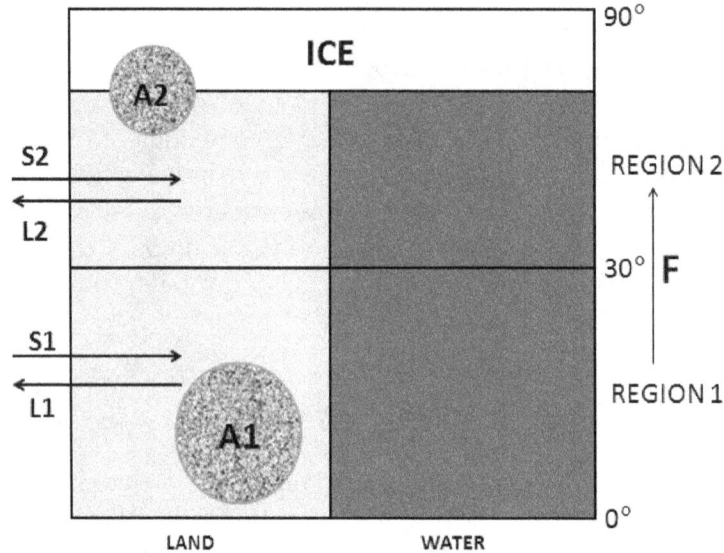

Figure 1: Schematics of the model. $S1$ and $S2$ are incoming shortwave fluxes. $L1$ and $L2$ are outgoing longwave radiation. F shows the poleward transfer of heat (latent and sensible) from the tropics to extra-tropics. Forested regions are depicted by $A1$ and $A2$.

the carbon forcing. Units for these parameters are in petawatts where 1 PW in either box is equivalent to $10^{15}/\pi r^2 = 7.8 \text{ W/m}^2$ at the top of the atmosphere. The atmosphere is assumed to have minimal heat capacity as compared to the land and ocean. In any case, for purposes of evaluation of the model, H determines only the relative time scale of the model response, hence precision in the actual value is completely unnecessary.

The snow/ice area as a fraction of the hemisphere is determined geometrically as mentioned above:

$$a = 1 - \sin\left(30° + 30°\frac{T_1 - T_{\text{ice}}}{T_1 - T_2}\right). \tag{2}$$

Albedo effects due to afforestation are incorporated into both the solar flux terms, S_1 and S_2, as well as effective ice albedo, α. Forest area is parametrized as a forest fraction f_i, which represents the fraction of the land surface of region i that is forested. Ice-free surface albedo is then calculated by an area-weighted average of ocean, forest, and barren land albedo. This surface albedo is then used to scale the incoming solar flux. Hence, both S_1 and S_2 depend, respectively, on independent forest fractions f_1 and f_2.

To determine the effect of ice on the radiation budget we first assume that the extra-topical forest is well-distributed throughout the region. Secondly, where forest and snow/ice-covered regions overlap we take the albedo to be that of the darker forest. These are both somewhat unrealistic assumptions, but they have been made in order not to underestimate the effect of extra-tropical afforestation on ice albedo. In fact, this assumption will very likely overestimate the albedo effect of extra-tropical afforestation since such efforts tend to occur in the mid-latitudes and the albedo of a snow-covered forested region will be necessarily be a value strictly between that of snow and forest (Betts and Ball 1997).

The ice albedo, $\alpha_I(f)$, a function of extra-tropical forest fraction, is then converted to an effective ice albedo which describes the net reflective effect of the ice-covered surface over that of the ice-free surface, $\alpha_L(f)$:

$$\alpha(f) = \frac{\alpha_I(f) - \alpha_L(f)}{1 - \alpha_L(f)}. \tag{3}$$

The atmospheric heat transport F is parameterized as follows:

$$F = F_0 + \gamma_1(T_1 - T_2) + \gamma_2 C(T_1)(T_1 - T_2)$$
$$C(T_1) = 6.11 \exp\left(17.23 \frac{T_1 - 273.15}{T_1 - 35.86}\right) \tag{4}$$

The first term in this formula for F describes the mean background value; the second and third terms are included to mimic the sensible and latent heat transports, respectively. Exponential dependence of latent heat transport on T_1 describes the moisture availability in the atmosphere. The particular form given here is that of the Magnus-Tetens approximation to the Clausius-Clapeyron equation which takes into account the temperature dependence of the latent heat of vaporization at the phase-change boundary. We assume that the majority of the moisture in the extra-tropical atmosphere comes from the tropics and therefore T_2 is not included in the expression for $C(T_1)$. Additionally, as discussed above, we relate moisture availability in the atmosphere to the tropical forest fraction, f_1, by scaling γ_2 by $(1 + f_1)/2$, which represents the fraction of the tropical box that is 'wet.'

Lastly, we assume the radiative forcing due to carbon sequestration in forests is uniform across both regions due to the fact that CO_2 is a well mixed greenhouse gas. This TOA forcing is computed via the relation given in REF:

$$\varepsilon = c \log\left(\frac{C_0 - C_1 - C_2}{C_0}\right) \tag{5}$$

where C_0 is a base atmospheric CO_2 value and $C_i = (\mu_i * 0.5 * 3.67)f_i L$. That is, C_i expresses the total CO_2 sequestered in a forest of area $f_i L$ where L is the area of each land region in hectares. The biomass density of the forest is μ (tonnes per hectare) and we assume that half of the forest biomass is carbon (Myneni et al. 2001, Penman et al. 2003). Carbon content is converted to CO_2 by multiplying by 3.67, which is the ratio of the molecular weight of CO_2 to that of carbon. The scaling parameter c in formula (5) is often taken to be 6.3 (see Table 2.2 in Houghton et al. 1990), but Myhre et al. (1998) use a detailed analyses of three broadband radiative transfer models to argue for a value of $c = 5.35$. It is this latter value which we use here.

The values of μ used in our simulations below are approximate values based on available data and research. As we noted in the introduction, estimation of forest carbon stocks is a complex field in itself. Given that each plantation is influenced by location-specific factors such as soil, hydrology, and microclimate, it is almost impossible to come up with generic formulae for a particular species. Though detailed procedures and estimates for volume, biomass and carbon content of different species have been produced for IPCC reports (e.g., Annex 3A.1 and 4A.2, Penman et al. 2003), extensive field research being carried out in different parts of the world has shown varying results.

For instance, Gonzalez et al. (2010) estimated carbon densities of forests in California, USA using remote sensing technologies, including lidar and satellite imagery, with calibration by in situ measurements. They report that both lidar and satelite image analysis produce lower estimates of forest carbon density than field estimates. They conclude that lidar captures a more complete picture of areas of low tree density than the field sample, whereas satellite image analysis seemed to systematically undercount live tree density.

Many estimates of forest biomass/carbon density have been produced. Brown and Lugo (1984) surveyed existing volume estimates of tropical forests and produced an estimated weighted biomass densities for undisturbed closed and open tropical broadleaf forests of 176 and 61 tonnes per hectare, respectively. However, a later study using more varied data sources (Brown and Lugo, 1992) saw them raise their mean biomass estimate to 300 T/ha. Milne and Brown (1997) combined numerous surveys and census data to estimate forest carbon densities in Great Britain. They estimated maximum stand densities for the oldest forests at 127 and 173 tonnes carbon per hectare for coniferous and broadleaf species, respectively. However, the mean densities (averaged over age and area distribution) were reported to be 21 and 61 tonnes/ha, respectively.

Similarly, Chaturvedi et al. (2011) report that tropical forests in India show a wide distribution of carbon density based on their age and location. For a very productive site, their sample shows a carbon density of 151 tonnes/ha with a growth rate of 5.3 tonnes/ha/year while on a less productive site the carbon density is reported to be 15.6 tonnes/ha with a growth rate of 0.05 tonnes/ha/year.

For our model, we use forest biomass densities that are consistent with the above estimates. Again, precision in these values is not essential in a qualitative study such as this where we seek to compare bulk climate effects of afforestation across regions.

4. Model Climates

We establish a base climatology range by first tuning our model to reproduce a present-day climate and then finding the equilibrium response of this model climate to uniform forcing (Figure 2). By a 'present-day' climate we mean one with average temperature, ice area, and heat transport at 30°N near to their present-day values. Of course, this is somewhat artificial given that we initialize our model with forest fractions prescribed to be zero. That is, our base climatology is determined by an Earth with barren land surface. Nevertheless, we do not consider this a problem given that (1) we are confining our analysis to a qualitative description of model differences with respect to various afforestation regimes and (2) our model is already highly idealized and is not meant to be used for either replication of current climate or prediction of future climates.

Figure 3 shows total AHT and ice latitude across a range of climates. We note that AHT shows a generally positive relation to global temperature with a saturation, and even a possible decrease, in the low-gradient, high-temperature regime located near the +2 PW forcing value. This is consistent with AHT behavior derived from more sophisticated models, for example Caballero and Langen (2005). However, for even warmer climates (upwards of +2 PW) AHT shows a renewed and steady upwards trajectory. This could lead to some concern that our simple parametrization of AHT in equation (4) is insufficient to describe AHT response in very warm climates. However, we note that the experiments carried out in this paper will

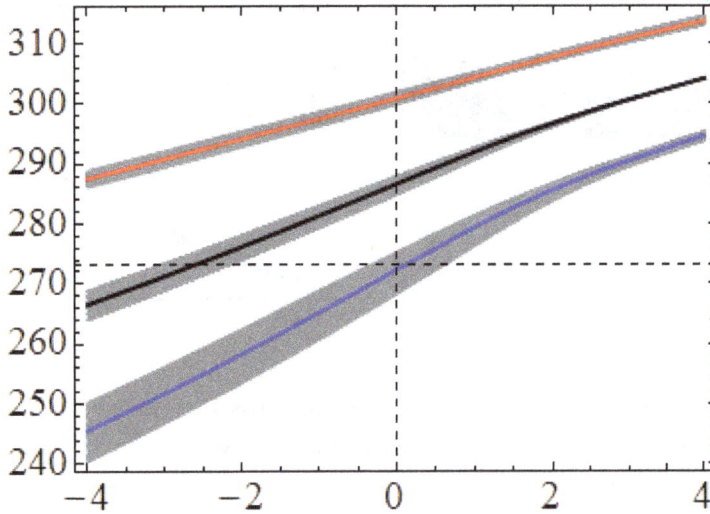

Figure 2: Model climates with $f_1 = f_2 = 0$. Horizontal axis shows TOA forcing value while vertical axis shows temperature. Model parameters are: $S_1 = 41$ PW; $S_2 = 24$ PW; $\alpha = 0.33$; $A = -49.0$ PW; $B = 0.29$ PW/K; $F_0 = 3$ PW; $\gamma_1 = 0.025$; $\gamma_2 = 0.0015$. Grey bands show the effect of 10 % variation in the values for effective ice albedo, γ_1, and γ_2.

never result in such high temperatures so we don't consider it a major cause for concern.

5. Afforestation Experiments

As discussed above, the model used in this study has been adapted from a model used to assess the relative roles of AHT and SAF in shaping the polar amplified response of the global climate to uniform forcing. But it is important to note that despite a more or less uniform impact on global CO_2 levels, the climate impact of afforestation is essentially non-uniform given its local effect on albedo and evapotranspiration. Nevertheless, these local effects can have large impacts across the globe due to the atmospheric circulation. Alexeev et al. (2005) investigated the effect non-uniform forcing on a 3D aquaplanet GCM. They found that even without ice-albedo feedbacks a +4 W/m2 forcing applied in the tropics resulted in a more or less uniform global response.

Figure 4 shows the outcome of our afforestation experiments. In these experiments the model was subjected to both tropical (case 'T') and extra-tropical (case 'X') afforestation and allowed to equilibrate. Forest fractions from 0 to 0.3 were prescribed in both regions in separate model runs. Three test cases were used corresponding to forest biomass densities of 100, 150, and 200 tonnes/ha. We refer to these cases as 'low,' 'medium,' and 'high' biomass density. Forrest albedo was kept fixed in all model runs.

One first notices that model temperatures decrease in all regions for medium to high density forestation, but show no change, or even warming, for low density forest. Of course, there is no difference here between low, medium, and high biomass forests in terms of their effect on surface albedo or evapotranspiration. Hence, the region-specific differences between temperature response in each case is due mainly to the difference in carbon uptake.

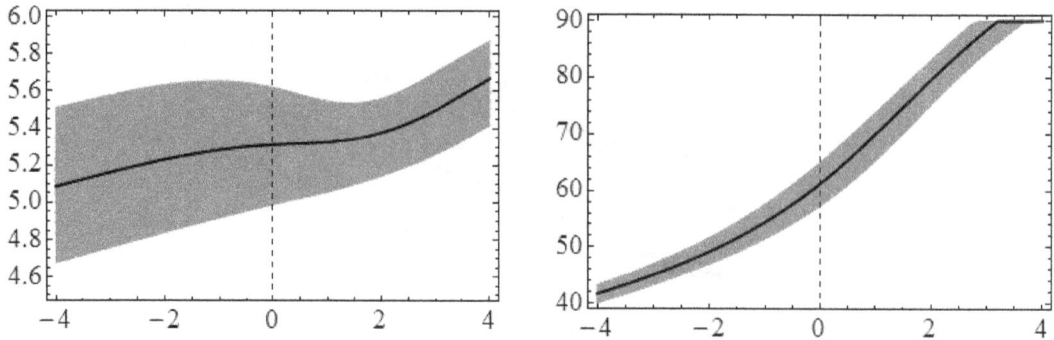

Figure 3: Total AHT in PW (left) and ice latitude in degrees (right) across a range of climates. Horizontal axis shows TOA forcing value. Grey bands show the effect of 10% variation in the values for effective ice albedo, γ_1, and γ_2.

Looking at tropical temperatures (Figure 4(a)) we see that the model response differs significantly depending on the region subjected to afforestation, i.e., case T or case X. In particular, afforestation in case X is seen to result in much cooler tropical temperatures when compared to equivalent afforestation in case T. Extra-tropical temperatures (Figure 4(b)) show a similar relation: afforestation in case T leads to cooler temperatures in the extra-tropics as compared to equivalent afforestation in case X. However, the magnitude of the temperature differences in the extra-tropics different afforestation regimes (T, X) are not as great under as they are in the tropics. These temperature differences are shown in Figure 4(e). We see, for instance, that for a forest fraction of 0.3 the tropics will be 2 K warmer if the afforested region is chosen to be in the tropics (case T) rather than in the extra-tropics (case X). On the other hand, the extra-tropics will be 1 K cooler in case T than in case X for the same forest fraction.

We conclude that tropical afforestation leads to increased meridional temperature gradients in our model (Figure 4(f)). This increase in gradient is accompanied by an increase in total AHT at 30°N (Figure 4(d)) which is remarkably consistent for all choices of forest density. Given that AHT acts to extract heat from the tropics, we conclude that the increased gradients in case T are due primarily to the surface albedo effect, both that of the forest and the ice cap: tropical afforestation lowers the albedo of the forested region which will lead to a significant increase in absorbed solar radiation, thus increasing the radiation budget in the tropics while, at the same time, ice albedo feedback tends to increase the meridional gradient in cooling scenarios due to the natural polar amplification exhibited by the model.

Extra-tropical afforestation will lead to decreased meridional temperature gradients, however, as can be seen in Figure 4(f). This is interesting because, again, in cooling scenarios one naturally expects to see polar amplification, and thereby an increase in gradient. This then is an artifact of the non-uniform forcing caused by regional afforestation. The natural polar amplified response of the model is overwhelmed by the local albedo effects of afforestation in case X.

It is interesting to note that the differences between case T and case X are much reduced when viewed from the perspective of global temperatures. Figure 4(c) shows the global temperature response as a function of forest fraction, while Figure 4(e) shows the difference in mean global temperature response between case T and case X. From these plots we see that in the simplest

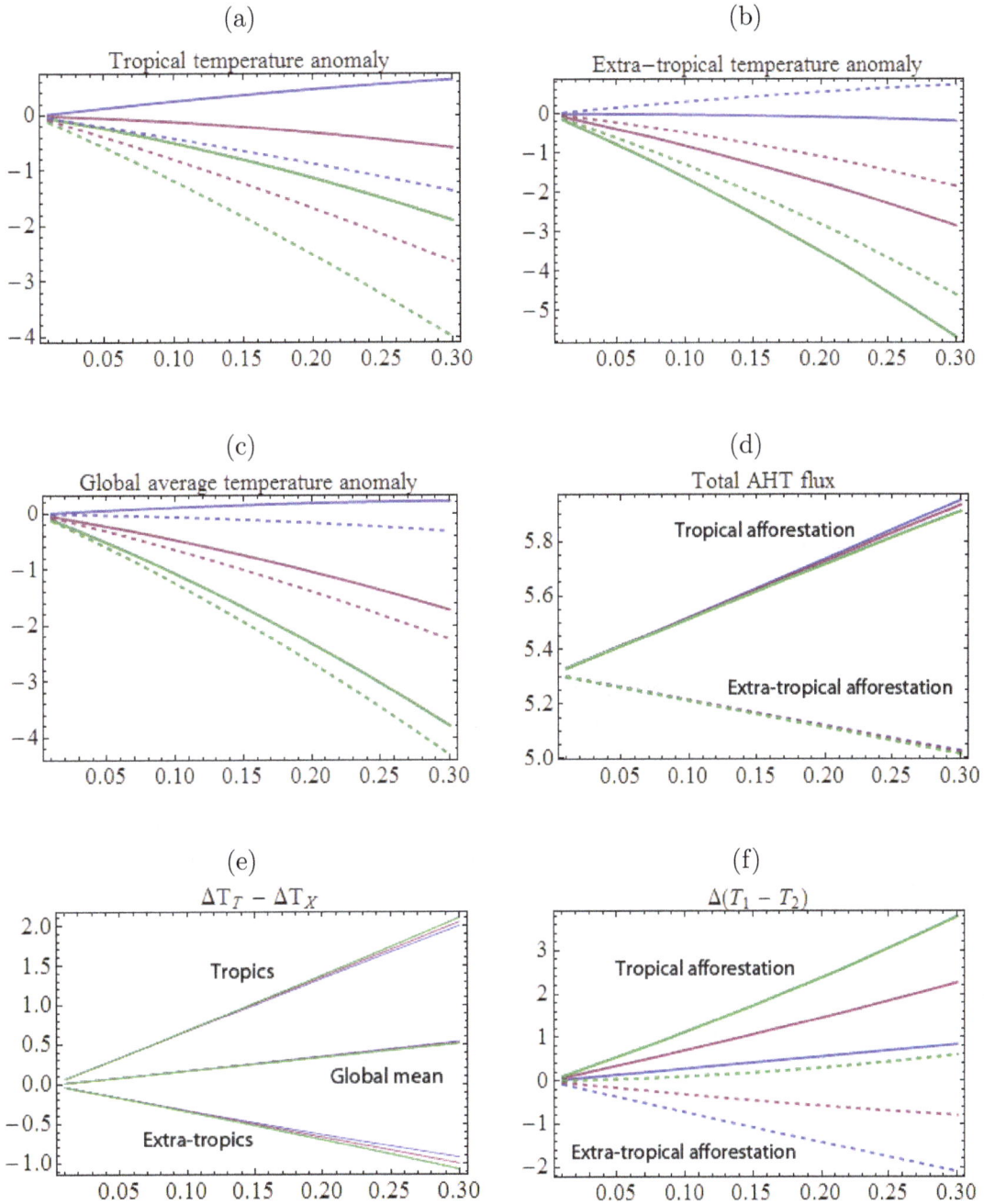

Figure 4: Results of afforestation experiments. Solid lines refer to tropical afforestation (case T), unless otherwise indicated. Dashed lines refer to extra-tropical afforestation (case X).

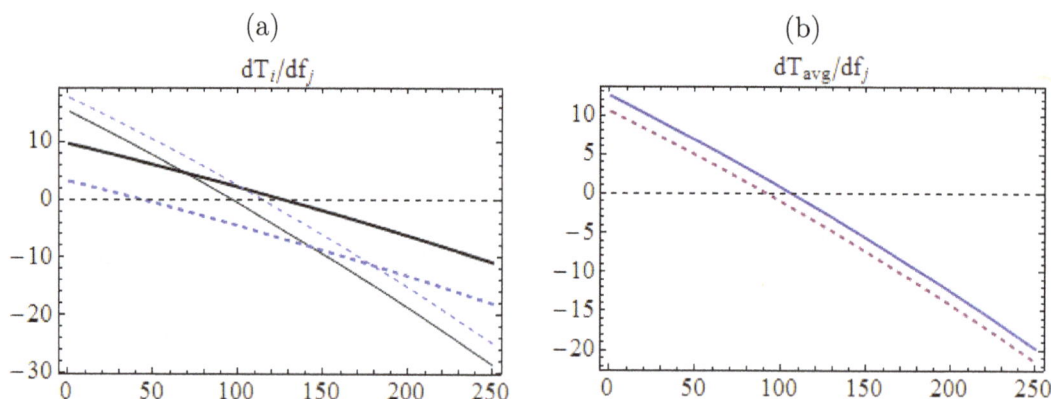

Figure 5: (a) dT_i/df_j for $i, j = 1, 2$; solid lines indicate a tropical afforestation scenario (case T) while dashed lines indicate extra-tropical afforestation (case X); thick lines correspond to tropical temperatures; dashed lines correspond to extra-tropical temperatures. (b) dT_{avg}/df_j for $j = 1, 2$; solid line indicates case T; dashed line indicates case X.

view of global temperature change, there is not much difference between afforestation in the tropics vs. the extra-tropics. Though, interestingly, what difference there is indicates that, all else being equal, a greater cooling effect will result from afforestation efforts carried out in the extra-tropics.

The fact that the response curves in Figure 4 are all linear suggests that for most model variables X, the rates dX/df_1 and dX/df_2 will depend only on the forest biomass density μ. At least, this is the case for sufficiently small forest fractions ($f_i \leq 0.3$). Figure 5 gives plots of these rates (as functions of biomass density, μ) for model temperatures, both globally and in individual boxes. In fact, nonlinearities do show up for very large forest fractions ($f_i > 0.8$) due both to nonlinear model processes (SAF, Clausius-Clayperion) as well as the logarithmic parametrization of the carbon forcing in equation (5).

The solid lines in Figure 5(a) correspond to a tropical afforestation scenario. The μ-value at which they cross (approx. 60 T/ha) is the forest biomass density for which tropical afforestation will result in uniform temperature change in both regions (warming in this case). For larger biomass densities tropical afforestation will result in increased temperature gradients and increased heat/moisture transport even in cooling scenarios (Figure 4(d)).

For extra-tropical afforestation (Figure 5(a) dashed lines), this point of intersection occurs for a much larger μ-value (approx. 180 T/ha). Hence, for most forest biomass densities in our test range (100-200 T/ha), extra-tropical afforestation results in decreased gradients and reduced heat/moisture transport.

It is worth noting once again that, as far as global temperatures are concerned, extra-tropical afforestation has the larger cooling effect in this model. This is seen in Figure 5(b) where the curve in case X crosses the axis first. The difference between mean temperatures in case T and case X are not as large as for regional temperature differences in these two scenarios. However, this difference is not insubstantial either. For instance, the μ-intercepts of the curves in Fig 5(b) differ by approximately 15 T/ha, or approximately 10% of our mean carbon density.

Such an amount could not be considered negligible when assessing the carbon sequestration potential of a given forestation project. Hence, it should not be ignored here where this difference represents the separation between overall warming and overall cooling of otherwise identical afforestation projects based on the region (tropical or extra-tropical) in which they are located.

6. Discussion and Conclusion

Carbon sequestration is only one means by which large scale afforestation projects impact the global climate system. In this study we used a simple 2-box model to illustrate the role of non-carbon processes (albedo effects, increase surface vapor flux, and atmospheric transport of latent and sensible heat) in shaping the global response to non-uniform forcing induced by afforestation.

Our model shows that tropical afforestation tends to increase meridional temperature gradients while extra-tropical afforestation tends to suppress them. Global mean temperatures in our model show a smaller dependence on the latitude of the afforested region, with high latitude plantations resulting in more global cooling than tropical plantations of the same size. This may seem somewhat surprising given that claims in the literature tend to suggest the opposite. However, tropical forests tend to have greater carbon densities and higher albedo than extra-tropical forests (Betts and Ball 1997; Culf et al. 1995). Hence, a comparative analysis such as ours which holds all parameters in common except latitude cannot treat this issue.

Complex general circulation models are growing in their predictive capacity. However, disagreement exists over the parameters needed for accurate quantification of particular afforestation activities as well as the best technologies for determining these parameters. Given this, and the high degree of variability of GCMs on relevant timescales, we feel that a qualitative approach using a simple model is a good alternative.

In any case, given that proponents of large scale afforestation have targets in the range of 30 million ha per year (Nilsson and Wolfgang 1995), it is important to consider more than just carbon when addressing the climate mitigation efficacy of afforestation.

References

[1] Alexeev VA (2003). Sensitivity to CO2 doubling of an atmospheric GCM coupled to an oceanic mixed layer: A linear analysis. Clim Dyn 20:775–787

[2] Alexeev VA, Langen PL, Bates JR (2005). Polar amplification of surface warming on an aquaplanet in "ghost forcing" experiments without sea ice feedbacks. Clim Dyn 24:655–666

[3] Alexeev A, Jackson C (2013). Polar Amplification: Is atmospheric heat transport important? Clim Dyn 41:533-547.

[4] Betts AK, Ball JH (1997). Albedo over the boreal forest. J. Geophys. Res. 102(24): 28,901–28,909

[5] Betts RA (2000). Offset of the potential carbon sink from boreal forestation by decreases in surface albedo. Nature, 408: 187–190.

[6] Bonan GB, Pollard D, Thompson SL (1992). Effects of boreal forest vegetation on global climate. Nature, 359: 716–718.

[7] Bonan G (2008). Forests and Climate Change: Forcings, Feedbacks, and the Climate Benefits of Forests. Science, 320: 1444–1449.

[8] Brown S, and Lugo A (1984). Biomass of Tropical Forests: A New Estimate Based on Forest Volumes. Science, 223 (4642): 1290–1293.

[9] Brown S, Lugo A (1992). Aboveground Biomass Estimates for Tropical Moist Forests of the Brazilian Amazon. Interciencia 17(1): 8–18.

[10] Caballero R, Langen PL (2005). The dynamic range of poleward energy transport in an atmospheric general circulation model. Geophys Res Lett, 32: L02705.

[11] Cai M (2005). Dynamical amplification of polar warming. Geophys Res Lett, 32:L22710.

[12] Cess R (1978). Biosphere-Albedo Feedback and Climate Modelling. Journal of the Atmospheric Sciences, 35 (9): 1765–1768.

[13] Chaturvedi RK, Raghubanshi AS, Singh JS (2011). Carbon density and accumulation in woody species of tropical dry forest in India. Forest Ecology and Management, 262 (8): 1576–1588.

[14] Christie SI, Scholes RJ (1995). CARBON STORAGE IN EUCALYPTUS AND PINE PLANTATIONS IN SOUTH AFRICA. Environmental Monitoring and Assessment, 38: 231–241.

[15] Culf AD, Fisch G, Hodnett MG (1995). The Albedo of Amazonian Forest and Ranch Land. J. Clim 8: 1544–1554.

[16] Dommenget D, Floter J (2011). Conceptual understanding of climate change with a globally resolved energy balance model. Clim Dyn, 37: 2143–2165.

[17] Flannery BP (1984). Energy-balance models incorporating transport of thermal and latent energy. J Atm Sci, 41:414–421.

[18] Foley JA, DeFries R, Asner GP, Barford C, Bonan G, Carpenter SR, Chapin FS, Coe MT, Daily GC, Gibbs HK, Helkowski JH, Holloway T, Howard EA, Kucharik CJ, Monfreda C, Patz JA, Prentice IC, Ramankutty N, Snyder PK (2005). Global Consequences of Land Use. Science, 309: 570–574.

[19] Gibbard S, Caldeira K, Bala G, Phillips TJ, Wickett, M (2005). Climate effects of global land cover change. Geophys. Res. Lett., 32: L23705.

[20] Gonzalez P, Asner G, Battles J, Lefsky M, Waring K, Palace M (2010). Forest carbon densities and uncertainties from Lidar, QuickBird, and field measurements in California. Remote Sensing of Environment, 114: 1561–1575.

[21] Graversen RG, Wang M (2009). Polar amplification in a coupled climate model with locked albedo. Clym Dyn 33:629–643.

[22] Hall A, Manabe S (1999). The Role of Water Vapor in Unpeturbed Climate Variability and Global Warming. Journal of Climate, 12.8: 2327–2346.

[23] Houghton RA (2005). Aboveground Forest Biomass and the Global Carbon Balance. Global Change Biology, 11: 945–958.

[24] Houghton JT, Jenkins GJ, Ephraums JJ (eds.) (1990). IPCC First Assessment Reports. Intergovernmental Panel on Climate Change. Cambridge University Press.

[25] Held IM, Brian SJ (2000). Water vapor feedback and global warming. ANNUAL REVIEW OF ENERGY AND THE ENVIRONMENT, 25: 441–475.

[26] Langen PL, Alexeev VA (2005). A study of non-local effects on the Budyko-Sellers infrared parameterization using atmospheric general circulation models. Tellus 57A:654–661.

[27] Langen PL, Alexeev VA (2007). Polar amplification as a preferred response in an idealized aquaplanet GCM. Clim Dyn, 29: 305–317

[28] Milne R, Brown TA (1997). Carbon in the Vegetation and Soils of Great Britain. Journal of Environmental Management, 49: 413–433.

[29] Myhre G, Highwood E, Shine K, Stordal F (1998). New Estimates of radiative forcing due to well mixed greenhouse gases. GEOPHYSICAL RESEARCH LETTERS, 25 (14): 2715–2718.

[30] Myneni RB, Dong J, Tucker CJ, Kaufmann RK, Kauppi PE, Liski J, Zhou L, Alexeyev V, Hughes MK (2001). A Large Carbon Sink in the Woody Biomass of Northern Forests. Proceedings of the National Academy of Sciences of the United States of America, 98 (26): 14784–14789.

[31] Nilsson S, Wolfgang S (1995). The carbon-sequestration potential of a global afforestation program. Climatic change, 30 (3): 267–293.

[32] Penman J, Gytarsky M, Hiraishi T, Krug T, Kruger D, Pipatti R, Buendia L, Miwa K, Ngara T, Tanabe K, Wagner F (eds.) (2003). Good Practice Guidance for Land Use, Land-Use Change and Forestry. Intergovernmental Panel on Climate Change. Institute for Global Environmental Strategies.

[33] Pielke RA, Marland G, Betts RA, Chase TN, Eastman JL, Niles JO, Niyogi DDS, Running SW (2002). The influence of land-use change and landscape dynamics on the climate system: relevance to climate-change policy beyond the radiative effect of greenhouse gases. Phil. Trans. R. Soc. Lond. A., 360: 1705–1719.

[34] Peixoto J, Oort A (1992). PHYSICS OF CLIMATE. Melville, NY: AIP Press.

[35] Rodgers KB, Lohmann G, Lorenz S, Schneider R, Henderson GM (2003). A tropical mechanism for Northern Hemisphere deglaciation. Geochem Geophys Geosys, 4(5):1046.

[36] Schneider EK, Lindzen RS, Kirtman BP (1997). A tropical influence on global climate. J Atmos Sci, 54:1349–1358.

[37] Soden BJ, Wetherald RT, Stenchikov GL, Robock A (2002). Global cooling after the eruption of Mount Pinatubo: A test of climate feedback by water vapor. Science, 296 (727): 727–730.

Affiliation:

Craig Jackson
Department of Mathematics and Computer Science
Ohio Wesleyan University
Delaware, OH 43015
E-mail: `chjackso@owu.edu`

Sriharsha Masabathula
Department of Economics
Ohio Wesleyan University
Delaware, OH 43015
E-mail: `ssmasaba@owu.edu`

3

Non-homogeneous Poisson process in the presence of one or more change-points: an application to air pollution data

Lorena Vicini
Federal University of Santa Maria

Luiz Koodi Hotta
University of Campinas

Jorge Alberto Achcar
Medical School of Ribeirão Preto

Abstract

We consider the problem of modeling the number of times that an air quality standard is exceeded in a certain period of time. We assume that the number of times the threshold is exceeded takes place according to a non-homogeneous Poisson process (NHPP) with the mean function modeled by the generalized gamma distribution. We consider models with and without change-points. When the presence of change-points is assumed, we have none , one, two or three change-points, depending on the data set. We use the Bayesian approach, where the posterior summaries of interest are obtained using standard Markov Chain Monte Carlo (MCMC) methods. We also discuss the use of different prior distributions for the parameters of the models, with an analysis of the convergence of the Gibbs sampling algorithm and sensitivity for the choice of different priors. To illustrate the proposed method we consider simulated data and a pollution data set from of a region of Mexico City.

Keywords: Bayesian analysis, Markov Chain Monte Carlo methods and simulation, multiple change-points, ozone air pollution.

1. Introduction

One problem that has affected many regions around the world is air pollution. In some places, such as in big cities and industrial regions, there is a higher concentration of pollution. However, due to wind, the air pollution can spread to other regions .

Air pollution has become a public health problem, since an increase in pollution can cause serious public health problems, such as diseases related to respiratory and cardiovascular systems; these have been highlighted in many health studies (see, for example, Braga et al., 2002 ; Gouveia et al., 2006). Air pollution is characterized by the presence of toxic gases and liquid or solid particles in the air. An important example of a pollutant is ozone, because when its concentration remains above a threshold level for a certain period of time, individuals exposed to it can suffer serious health problems (see, for example, Air Resource Board (ARB), 2005).

In this paper we analyze a series of data for ozone (O_3), which is a gas composed of three oxygen atoms and formed by chemical reactions between nitrogen oxides (NOx) and volatile organic compounds $(VOC's)$ in the presence of sunlight. Ozone has the same chemical structure miles above the earth or at ground level and it can be "good" or "bad" depending on its location in the atmosphere. In the lower atmosphere, the tropospheric ozone is considered "bad".

As pollution levels have increased at an alarming rate in recent years, exceeding the limits for acceptable standard of air quality on certain days, studies related to the problem are gaining prominence around the world. As a direct result, new models and statistical methods have been developed to analyze air pollution data. Considering as the main interest the estimation of the number of times that a given environmental standard is violated, Javits (1980) assumes Bernoulli and Poisson models; Raftery (1989) uses a mixture of homogeneous Poisson models. Since time homogeneity is usually not verified in applications to air pollution data some authors assume non-homogeneous Poisson processes. It is important to point out that even assuming non-homogeneous Poisson models, usually we could have the presence in the model of one or more change-points (see, for example, Achcar et al., 2010; Achcar et al., 2011).

In relation to pollution by ozone gas, the literature contains several studies (see, for example, Wilson et al., 1980; Loomis et al., 1996; Galizia and Kinney, 1999; Bell et al., 2004; Gauderman et al., 2004; Álvarez et al.; 2005, ARB, 2005; Bell et al., 2005; Bell et al., 2007 ; and Achcar et al., 2010).

Other studies are also related to diseases caused by an increase in the level of air pollution (see, for example, Martins et al., 2002; Farhat et al., 2005).

In this paper, we consider how to model the data by a non-homogeneous Poisson process (NHPP) with the rate function modeled by the generalized gamma distribution. The model is used to analyze the daily data set collected by the monitoring network of the Metropolitan Area of Mexico City. The set contains 18 years of daily average ozone measurements in the period from 1 January 1990 to 31 December 2008. The Metropolitan Area of Mexico City is divided into five regions, corresponding to the Center (CE), Northwest (NW), Northeast (NE), Southeast (SE) and Southwest (SW).

This paper considers the modeling of the data in the presence or not of one or more change-points. We get posterior summaries of interest using standard MCMC methods, in special,the Gibbs sampling algorithm or the Metropolis-Hastings algorithm. Also it is discussed the sensitivity of the choice of different prior distributions for the parameters of the model and their effect in the convergence of the simulation algorithm. The studies are illustrated with simulated data and a pollution real data set.

This article is organized as follows: in Section 2, the model is presented; Section 3 presents the Bayesian formulation, first without taking into account change-points, which are incorporated later; examples considering simulated data set are given in Section 4; in Section 5, the proposed models are applied to a data set collected by the monitoring network of the Metropolitan Area of Mexico City; Section 6 concludes with final remarks and discussions of the results.

2. Description of the model

The NHPP model has been used to model various phenomena. For example, times from remission for patients with leukemia (Matthews and Farewell, 1982), intervals between coal-mining disasters (Raftery and Akman, 1986, and Yang and Kuo, 2001), time of failures in repairable systems (Ruggeri and Sivaganesan, 2005), arrival times of calls to a call center (Weinberg et al., 2007). Other examples can be found in Leemis (1991). The problem can be described as follows. Let $T > 0$ be a real number and $M = \{M(t) : t \in (0, T]\}$, where the random variable $M(t)$ represents the cumulative number of events in the time interval $[0, t)$ for $t \geq 0$. In the NHPP model the random variable $M(t)$ has a Poisson distribution with mean $m(t)$. One can also characterize the distribution by the intensity function $\lambda(t) = \frac{dm(t)}{dt}$. If $\lambda(t)$ is a constant, so that $m(t)$ is linear, then $M(t)$ is called a homogeneous Poisson process; otherwise the process is called a non-homogeneous Poisson process.

Different choices for the function $m(t)$ are considered in the literature, especially in software reliability modeling (see, for example, Achcar et al., 1998). Goel and Okumoto (1979) stated that the expected number of software failures for time t is given by the mean value function $m(t)$, which is non-decreasing and bounded above. Specifically, they considered the mean and intensity functions given, respectively, by,

$$m_1(t) = \theta \left(1 - e^{-\beta t}\right), \quad \text{and} \tag{1}$$

$$\lambda_1(t) = \theta \beta e^{-\beta t}, \tag{2}$$

where, in our case, θ represents the expected maximum number of days in which the air quality standard is violated by a particular pollutant and β is considered to be the rate at which events occur. Goel (1983) generalized model (1) proposing the intensity function

$$\lambda_2(t) = \theta \beta \alpha t^{\alpha-1} e^{-\beta t^\alpha}, \tag{3}$$

with mean function

$$m_2(t) = \theta \left(1 - e^{-\beta t^\alpha}\right). \tag{4}$$

Note that (1) and (4) can be written as special cases of the general form where the mean value function is given as

$$m(t) = \theta F(t), \tag{5}$$

where $F(t)$ is a distribution function. On the other hand, for any distribution function $F(t)$ we have a valid model.

A widely used distribution function, given its high flexibility, is the generalized gamma distribution. In this case the mean function is

$$m_{GG}(t) = \theta I_k(\beta t^\alpha), \tag{6}$$

where $I_k(s)$ is the integral of the gamma function given by

$$I_k(s) = \frac{1}{\Gamma(k)} \int_0^s x^{k-1} e^{-x} dx. \tag{7}$$

From (7), we obtain the intensity function given by

$$\lambda_{GG}(s) = m'_{GG}(t) = \frac{1}{\Gamma(k)} \theta \beta^k \alpha t^{\alpha k - 1} e^{-\beta t^\alpha}. \tag{8}$$

This model is called generalized gamma. When k is an integer we can write $F(t)$ as

$$F(t) = 1 - e^{-\beta t^\alpha} \sum_{j=o}^{k-1} \frac{(\beta t^\alpha)^j}{j!}. \tag{9}$$

The three models used in this article are described as follows:
a) Model I: Here all parameters are unknown, $\lambda_{GG}(t)$ is given by (8).
b) Model II: With $k = 1$, $\lambda_{GG}(t)$ is given by $\lambda_2(t)$ (3).
c) Model III: With $k = 1$ and $\alpha = 1$, $\lambda_{GG}(t)$ reduces to $\lambda_1(t)$ (2).

In the Bayesian analysis of these models, we may have some difficulties in obtaining the Bayesian inferences using MCMC simulation methods. These difficulties are mainly related to the convergence of the MCMC chains. For this reason we will study the effect of different prior distributions on the performance of sample simulation algorithms of the posterior distribution of interest.

The main idea of this paper is to model the number of times that an air quality standard is exceeded in a period of time, under a Bayesian approach. To do this, we explore the generalized intensity functions that can best fit the data set. In particular, we fit the generalized intensity function with and without the presence of change-points. We also study several prior distributions for the parameters because of the convergence problem mentioned above, especially when all the parameters are estimated simultaneously. For this reason we perform sensitivity analysis in relation to the choice of prior distributions. The biggest concern is related to the parameter k, a parameter of the gamma distribution.

The convergence of the MCMC algorithm was analyzed by graphical methods and by the Gelman-Rubin statistic (Gelman and Rubin, 1992). The Gelman-Rubin statistic relies on parallel chains to test whether they all converge to the same posterior distribution. Brooks and Gelman (1998) suggested the introduction of a correction factor. This statistic is evaluated using the coda package in R. This statistic is larger or equal to 1.0. The closest this statistic is to 1.0, more evidence we have that the chain is near convergence. The limit value of 1.2 is sometimes used as a guideline for "approximate convergence" (Gelman, 1996).

3. Bayesian inference

Denote by $D_T = \{n; t_1, t_2, ..., t_n; T\}$, the data set, where n is the number of events observed such that $0 \leq t_1 < t_2 < ... < t_n \leq T$, and where t_i are the times of the events observed during the period of time $(0, T]$.

We consider that the parameters θ, β, α and k are unknown and will be estimated. In the Bayesian framework, for each parameter, we must select prior distributions which describe the uncertainty about them.

In this article we consider the presence or absence of change-points to NHPP. We have two different forms for the likelihood function of the model, one for each formulation. First, we define the notation and expressions for the case without change-points, and then we do the same for the case when there are one or more change-points.

3.1. Models without change-points

For Model I, the likelihood function for the vector $\Theta = (\theta, \beta, \alpha, k)$, considering T as the truncation time of the truncated model (see, for example, Cox and Lewis, 1966), is given by

$$L(\Theta \mid D_T) = \left(\prod_{i=1}^{n} \lambda(t_i) \right) \exp(-m(T)). \tag{10}$$

In some cases it is advisable to enter a latent variable, as this may serve as a computational aid. We consider the introduction of the latent variable N' which has Poisson distribution with parameter $\theta[1 - F(T \mid \beta)]$ (see, for example, Achcar et al., 1998).

Considering the generalized gamma distribution model given in (6), the likelihood function for the vector of parameters $\Theta = (\theta, \beta, \alpha, k)$ is expressed as

$$L(\Theta \mid D_T) = \left\{ \prod_{i=1}^{n} \theta \frac{\beta^k}{\Gamma(k)} \alpha t_i^{\alpha k - 1} e^{-\beta t_i^\alpha} \right\} \exp \left\{ - \int_o^T \theta \frac{\beta^k}{\Gamma(k)} \alpha u^{\alpha k - 1} e^{-\beta u^\alpha} du \right\},$$

such that

$$L(\Theta \mid D_T) = \frac{\theta^n \alpha^n \beta^{kn}}{\{\Gamma(k)\}^n} \left\{ \prod_{i=1}^{n} t_i^{\alpha k - 1} \right\} \exp \left\{ -\beta \sum_{i=1}^{n} t_i^\alpha - \theta I_k(\beta T^\alpha) \right\}. \tag{11}$$

The prior distributions and the conditional posterior distributions of the MCMC algorithm are presented in the appendix.

3.2. Models with change-points

The ozone pollution often changes during the time interval $(0, T]$, due to some type of intervention or change. In special, some political decisions by public authorities could implies in a decreasing or an increasing in ozone gas emission and, therefore, a similar effect can occur in the daily measurements of ozone.

In this case we can have J change-points, which we denote by τ_j, $j = 1, ..., J$, and in each interval we use the generalized gamma model, the NHPP model, presented in Section 2.

We assume here that all the parameters are unknown: $(\theta_i, \alpha_i, \beta_i, k_i)$ and τ_j, where $i = 1, ..., J+1$ and $j = 1, ..., J$. They must be estimated, and the change-points can occur at any time $\tau_0, \tau_1, \cdots, \tau_i$, where $\tau_0 = 0$.

In this case, we have that the rate function of the NHPP process is of the form (Achcar et al., 2010)

$$\lambda(t \mid \Theta) = \begin{cases} \lambda(t \mid \Theta_1), & \text{se } 0 \leq t < \tau_1 \\ \lambda(t \mid \Theta_j), & \text{se } \tau_{j-1} \leq t < \tau_j, j = 2, 3, ..., J, \\ \lambda(t \mid \Theta_{J+1}), & \text{se } \tau_J \leq t \leq T, \end{cases} \tag{12}$$

where the intensity functions $\lambda(t \mid \Theta_j)$, $j = 1, 2, ..., J+1$ are related to equation (12) and $\Theta_j = (\theta_j, \beta_j, \alpha_j, k_j)$, $j = 1, 2, ..., J+1$ are the parameters associated with the NHPP process in each interval limited by the change-points.

For the generalized gamma distribution the rate function is of the form

$$\lambda(t \mid \Theta) = \frac{1}{\Gamma(k_j)} \theta_j \beta_j^{k_j} \alpha_j t^{\alpha_j k_j - 1} e^{-\beta_j t^{\alpha_j}}.$$

The mean function $m(t \mid \Theta_j), j = 1, 2, ..., J+1$, as given in Achcar et al. (2010) is of the form

$$m(t \mid \Theta) = \begin{cases} m(t \mid \Theta_1), & \text{se } 0 \le t < \tau_1 \\ m(\tau_1 \mid \Theta_1) + m(t \mid \Theta_2) - m(\tau_1 \mid \Theta_2), & \text{se } \tau_1 \le t < \tau_2, \\ m(t \mid \Theta_{j+1}) - m(\tau_j \mid \Theta_{j+1}) \\ \quad + \sum_{i=2}^{j} [m(\tau_i \mid \Theta_i) - m(\tau_{i-1} \mid \Theta_i)] + m(\tau_1 \mid \Theta_1), & \text{se } \tau_j \le t < \tau_{j+1}, \\ j = 2, 3, ..., J, \end{cases} \tag{13}$$

where $\tau_{J+1} = T$. For the generalized gamma distribution the mean function is given by

$$m(t \mid \Theta) = \theta_j I_{k_j}(\beta_j t^{\alpha_j}).$$

Let $\boldsymbol{w} = (\Theta_1, \Theta_2, ..., \Theta_{J+1}; \tau_1, \tau_2, ..., \tau_J)$ is the vector of all parameters. Having assumed an NHPP, the likelihood function is given by

$$L(w \mid D_T) \propto \prod_{i=1}^{N_{\tau_1}} \lambda(t_i \mid w_1) e^{-m(\tau_1 \mid w_1)} \left[\prod_{j=2}^{J} \left(\prod_{i=N_{\tau_{j-1}}+1}^{N_{\tau_j}} \lambda(t_i \mid w_j) e^{-[m(\tau_j \mid w_j) - m(\tau_{j-1} \mid w_j)]} \right) \right]$$

$$\times \prod_{i=N_{\tau_J}+1}^{N_T} \lambda(t_i \mid w_{J+1}) e^{-[m(T \mid w_{J+1}) - m(\tau_J \mid w_{J+1})]}. \tag{14}$$

The inclusion of change-points in the restricted Models II and III is carried out in the same way as in the inclusion in Model I.

4. Application to simulated data

We consider models without change-points and with one and two change-points. These models will initially be tested with simulated data sets and later they will be used to model ozone pollution data from Mexico City based on when a threshold in the levels of ozone concentration is exceeded for a certain period of time. In both cases we will discuss the convergence of the chains and the adjustment of the function $m(t)$. In most cases the number of simulated data is not large because we want to test whether the method works in this situation. Thus, we do not expect to find small variances for the posterior distributions.

4.1. Simulated data without change-point

We first simulated 300 observations with $\theta = 300$, $\beta = 0.02$, $\alpha = 1$ and $k = 1$, i.e. Model III without a change-point. Models I and II are also correct because they incorporate this model, but they are over-specified.

We obtained a summary of estimates of the posterior distributions of Model I based on $900,000$ MCMC replications after a burn-in of size $50,000$ and jumps of 500 samples (A jump of 500 means that every 500th sample is chosen from the simulated Gibbs samples to have approximately uncorrelated samples used to get Monte Carlo estimates for the random quantities of interest.) For the restricted Models II and III, summaries of the estimates of the posterior distributions were obtained using $100,000$ iterations, a burn-in of size $10,000$ and jumps of 10 samples.

As the analysis of the adjustment of the models without change-point was carried out in Vicini et al. (2012), we only present the fit of the three models and the summaries of the estimated posterior distributions. Vicini et al. (2012) analyzed the sensitivity of the estimates of all the parameters, and they found difficulties in obtaining convergence of the chains for the parameter k_i of the Model, when assuming the improper prior used in the first set of distributions. They tested various specifications of prior distributions and different values for hyperparameters and suggested the truncated exponential distribution, which is used here.

Figure 1 shows the adjustment of Models I, II and III in the absence of a change-point. This figure presents the graphs of the theoretical function $m(t)$, its empirical (or nonparametric) estimation, and its estimation by Models I, II and III without a change-point. We observe a good fit of all assumed models to the data set. We observe that the estimated curves are superimposed on each other and on the empirical curve. The curves are evaluated at the parameter values given by the median of the posterior distribution, since these estimates provide better results than using the mean values. From now on we will always use the posterior medians in place of the posterior means.

Table 1 presents the estimates of the mean, median and standard deviation of the parameters of Models I, II and III without change-points. The 95% credible intervals (CI) were obtained by taking the 2.5% and 97.5% percentiles of the simulated Gibbs sample. We observe that the parameter values were generally in their respective 95% CI. In this paper, we use the 2.5% and 97.5% percentiles for all the CI.

Table 1: Summary of the posterior distributions of the parameters for Models I, II and III, with no change-points. Real model:$\theta = 300$, $\beta = 0.02$, $\alpha = 1$ and $k = 1$. Gibbs sample of size 300. Prior distributions for the parameters θ, β and α are presented in Section (3). The prior distribution for the parameter k in Model I is the exponential distribution with mean parameter equal to 0.95 and truncated at 3.

Model	parameter	mean	median	S.D.	(CI)2.5%	(CI)97.5%
I	θ	310.00	309.47	17.86	282.30	340.10
	α	1.234	1.219	0.174	0.976	1.543
	β	0.007687	0.005503	0.007301	0.000941	0.022115
	k	0.765	0.751	0.161	0.528	1.057
II	θ	310.28	309.89	17.63	282.27	339.70
	α	1.002	1.002	0.049	0.922	1.085
	β	0.019282	0.018933	0.004088	0.013193	0.026475
III	θ	310.03	309.82	17.77	281.62	339.49
	β	0.019115	0.019087	0.001214	0.017144	0.021113

We observe in Table 1 that the Monte Carlo estimates for the posterior mean and posterior median of the parameters of the subsequent Models I, II and III were satisfactory, because their values are very close to the true values of the parameters. We can also check that all CI contain the true value of the parameters. We observed good convergence of the simulation

m(t)

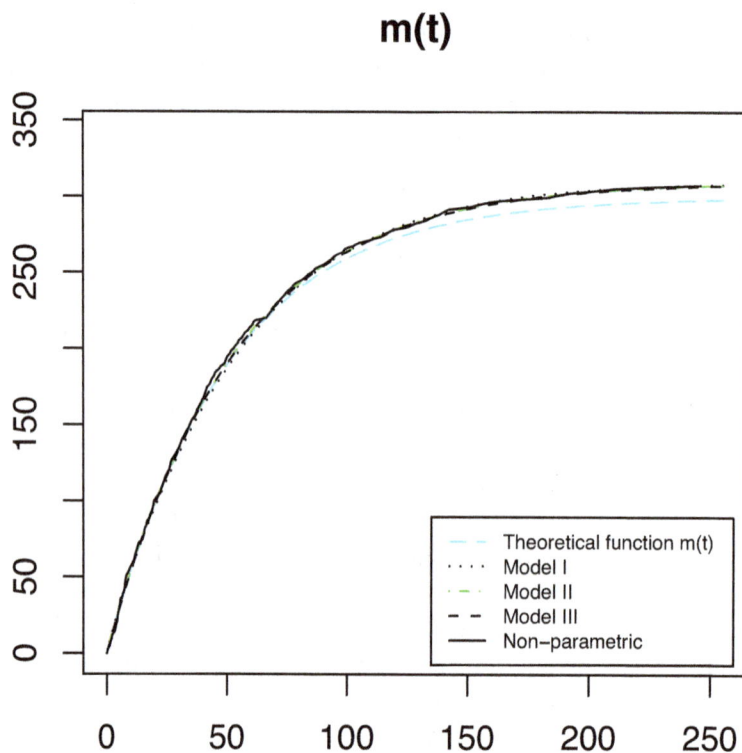

Figure 1: Simulation from the model without change-point with parameters $(k, \alpha) = (1, 1)$. Sample size equal to 300. Graphs of the theoretical function $m(t)$, the empirical estimation and its estimates by models without change-point: a) Model I, b) Model II c) Model III.

algorithm, as observed in the trace plots of the simulated samples. The largest value of the Gelman-Rubin statistic is equal to 1.01 also indicating convergence of the MCMC chains. The results indicate that the information that $\alpha = k = 1$ is not relevant to the estimation of the parameter θ, as the estimate of the posterior distribution remains almost the same whether or not the constraint is taken into account.

4.2. Simulated data with two change-points

In this case, we simulated 150 observations with parameter values given by $\theta_1 = 190$, $\theta_2 = 130$, $\theta_3 = 365$, $\beta_1 = 0.009$, $\beta_2 = 0.007$, $\beta_3 = 0.003$, $\alpha_1 = 1$, $\alpha_2 = 1$, $\alpha_3 = 1$, $\tau_1 = 50$, $\tau_2 = 100$, $k_1 = 1$, $k_2 = 1$ and $k_3 = 1$, with change-points $\tau_1 = 50$ and $\tau_2 = 100$, with 50 observations before τ_1, 50 observations between τ_1 and τ_2 and 50 after τ_2. The prior distributions for the parameters k_1 and k_2 in Model I are the second set of the proposed distributions presented in Section 3, where we have the exponential prior distribution with mean parameter equal to 0.99 and truncated at 6, for both parameters.

Here all three models, Models I, II and III, are correct, but Model II incorporates the information that $k_1 = k_2 = 1$ and Model III incorporates the information that $\alpha_1 = \alpha_2 = \alpha_3 = k_1 = k_2 = k_3 = 1$, while in Model I these parameters are estimated.

The posterior summaries of Model I were obtained using $100,000$ iterations, a burn-in of size $30,000$ and jumps of size 50 in the simulation algorithm. For the posterior summaries of Models II and III we used $80,000$ iterations, a burn-in of size $20,000$ and jumps of size 50.

We also performed a sensitivity analysis with respect to the specifications of the a priori distributions. Here we present only the best obtained parameter estimates for these models, which are given using the following specifications of prior distributions for the parameters: $\theta_1 \sim Gamma(0.001, 0.001)$; $\theta_2 \sim Gamma(0.001, 0.001)$; $\theta_3 \sim Gamma(0.001, 0.001)$; $\beta_1 \sim Gamma(1, 100)$; $\beta_2 \sim Gamma(1, 100)$; $\beta_3 \sim Gamma(0.01 \times 100, 100)$; $\alpha_1 \sim Gamma(1, 1)$; $\alpha_2 \sim Gamma(1, 1)$; $\alpha_3 \sim Gamma(1, 1)$; $\tau_1 \sim Uniform(40, 60)$ and $\tau_2 \sim Uniform(90, 110)$.

We only present the graphical results of Model I, where all the parameters are free.

Figure 2 shows the graphs of the chains for the parameters of Model I with two change-points. Figure 3 shows the posterior and prior distributions and the true value of each parameter for Model I with two change-points.

The analyses of Figures 2 and 3 give an indication that the chains are converging to their true values, as can be seen from the horizontal trace in Figure 2 and the vertical line in Figure 3.

Figure 4 shows the fit of Models I, II and III, with the presence of two change-points. This figure presents the graphs of the theoretical function $m(t)$, its empirical estimation, and its estimation by Models I, II and III. All models with two change-points give a good fit for the simulated data. This result can be seen from the curves that overlap with each other.

Table 2 presents the estimates for the parameters of Models I, II and III with two change-points. Analyzing Table 2 we can observe that the estimates of the mean and median of Models I, II and III with two change-points were satisfactory, because their values are very close to the true values of the parameters. All the credibility intervals 95% contain the true value of the parameters. There was no problem in the convergence of the chains of the parameters, as shown by the chains and also by the good fit of the curves. The largest value of the Gelman-Rubin statistic is equal to 1.15, obtained for the parameter θ_2, also indicating convergence of the MCMC chains. Note that the three models are correct, but Models II and III incorporate more information, and Model III incorporates more information than Model II. We also see, by comparing the results of Model III with the other models, that adding the information that $k_1 = k_2 = k_3$ changes very little the posterior distribution of the parameter β_1, indicating that information on the parameters $k's$ is not relevant to the parameter β_1. Simultaneously analyzing Tables 1 and 2, we can see that the range of the credibility intervals of the parameters are larger in Table 2, because the sample size is smaller in this case.

5. Application to ozone data

Air pollution is one of the main problems facing large cities. There are many pollutants that cause problems to the population, but what most affects large cities is ozone gas. In some cities, like Mexico City, the authorities have been concerned with the high levels of pollution that it presents. Related to this problem, the environmental authorities have implemented measures aimed at reducing the level of pollutants. These measures are extremely important, because when pollution levels reach a certain threshold of concentration for a given period of

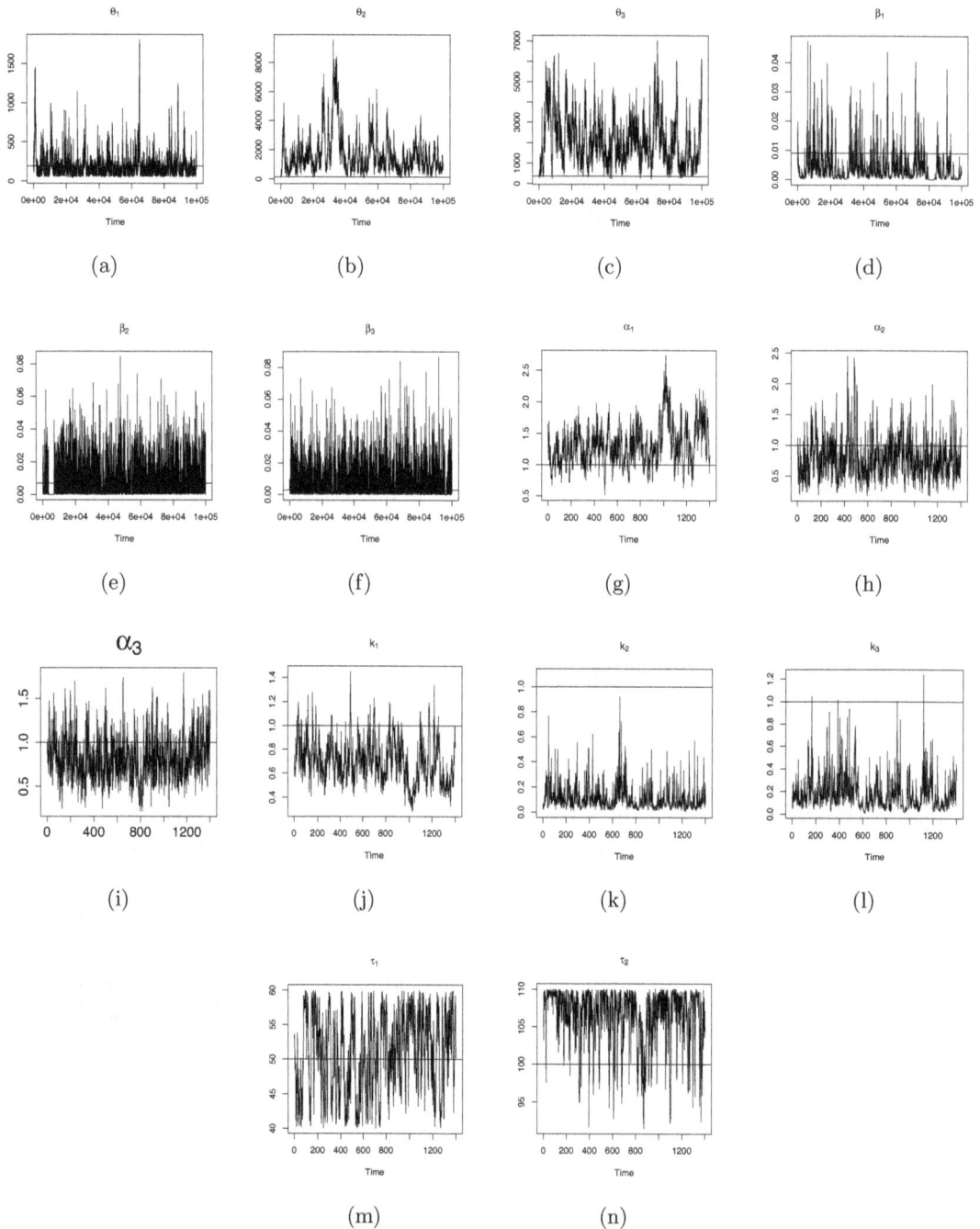

Figure 2: Simulation of the chains of Model I with two change-points. The horizontal traces are the true values of each parameter.

time, individuals exposed to the pollutant can suffer serious health problems (see, for example, Wilson et al., 1980; Loomis et al., 1996; Galizia and Kinney, 1999; Bell et al., 2004; Bell et al., 2005; Bell et al., 2007; Gauderman et al., 2004; and ARB, 2005). In this section, we apply the proposed models to the measurements of ozone in the Metropolitan Area of Mexico City. The Mexican ozone standard is 0.11 parts per million (0.11 ppm) and the threshold used in

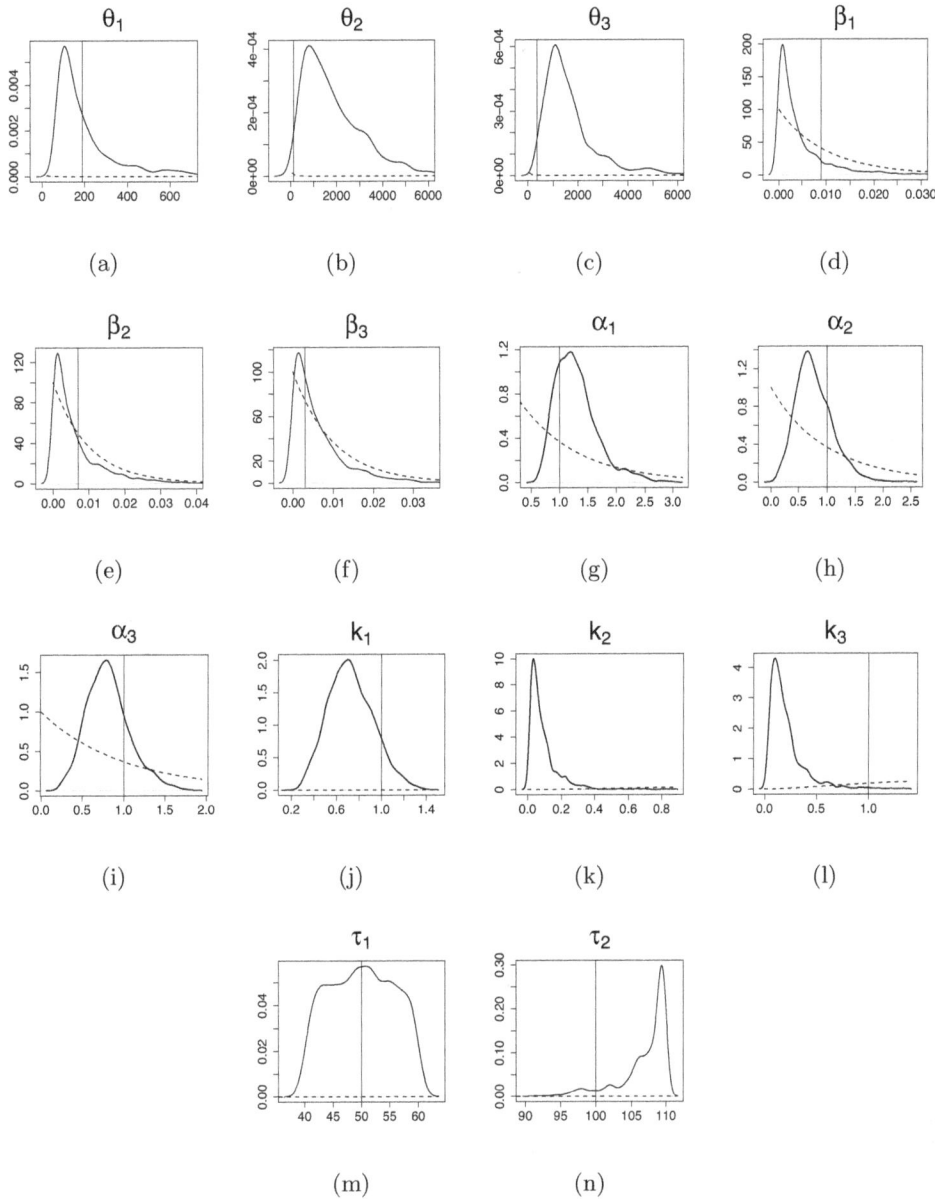

Figure 3: Simulation of the posterior distribution of the parameters of Model I with two change-points. The posterior distributions are shown with a solid line and the prior distributions are shown with the dashed line. The true values are indicated by vertical lines.

Mexico City to declare an emergency is 0.22 ppm (see, for example, Achcar et al., 2010). Here we consider a threshold equal to 0.20 ppm. This value is used because it is between the other two. We applied the proposed models to fit the data corresponding to the maximum daily average measurements of ozone gas, based on data measured in the Northeast region (NE) of Mexico City with a sample of 981 observations, which correspond to times when a certain threshold established for the air quality standard is violated during the period of time T which we considered. These data are collected from www.sma.df.gob.mx/simat/, which account for about 18 years of observations (1 January 1990 to 31 December 2008).

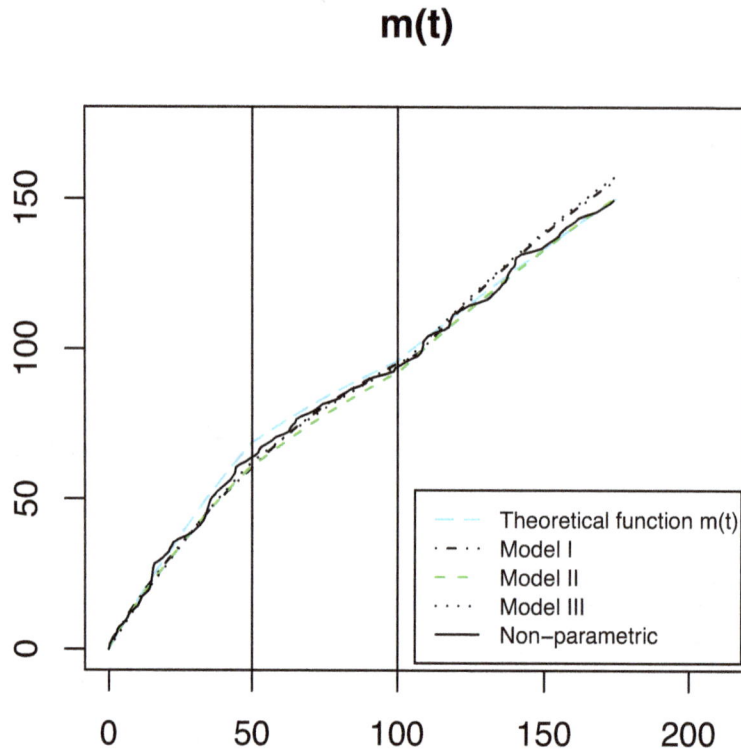

Figure 4: Model with two change-points. Sample size equal to 150. Graphs of the theoretical function $m(t)$, the empirical estimation and its estimates by models with two change-points: a) Model I, b) Model II c) Model III. The vertical lines indicate the change-points.

The posterior summaries of interest for Model I with no change-points were obtained using $700,000$ iterations, a burn-in of size $15,000$ and jumps of size 100. The specification of the prior distributions of the parameters for this model were: $\theta \sim Gamma(0.001, 0.001)$; $\beta \sim Gamma(10, 100)$; $\alpha \sim Gamma(0.01, 0.01)$. The assumed prior distribution for the parameter k is the exponential distribution with the mean parameter equal to 0.95 and truncated at 3.

For Model I with one change-point, the posterior estimations are obtained using $400,000$ iterations, a burn-in of size $200,000$ and jumps of size 100. The specifications of the prior distributions of the parameters for this model were: $\theta_1 \sim Gamma(0.001, 0.001)$; $\theta_2 \sim Gamma(0.001, 0.001)$; $\beta_1 \sim Gamma(0.001, 0.001)$; $\beta_2 \sim Gamma(0.001, 0.001)$; $\alpha_1 \sim Gamma(1, 1)$; $\alpha_2 \sim Gamma(1, 1)$ and $\tau_1 \sim Uniform(2900, 2950)$. The assumed prior distributions for the parameters k_1 and k_2 are the exponential distribution with mean parameter 0.99 and truncated at 6 for both parameters.

For Model I with two change-points the posterior estimations were obtained using $400,000$ iterations, a burn-in of $300,000$ and jumps of 200. The specifications of the prior distributions of the parameters for this model were $\theta_1 \sim Gamma(0.001, 0.001)$; $\theta_2 \sim Gamma(0.001, 0.001)$;

Table 2: Summary of estimates of the posterior distributions for Models I, II and III with two change-points. Real model: $\theta_1 = 190$, $\theta_2 = 130$, $\theta_3 = 365$, $\beta_1 = 0.009$, $\beta_2 = 0.007$, $\beta_3 = 0.003$, $\alpha_1 = 1$, $\alpha_2 = 1$, $\alpha_3 = 1$, $\tau_1 = 50$, $\tau_2 = 100$, $k_1 = 1$, $k_2 = 1$ and $k_3 = 1$. Sample size equal to 150. Prior distributions for the parameters θ_1, θ_2, β_1, β_2, α_1, α_2 and τ, are presented in Section 3. The prior distributions for the parameters k_1 and k_2 in Model I are the exponential distribution with the mean parameter equal to 0.99 and truncated at 6, for both parameters.

Model	parameter	mean	median	S.D.	2.5%	97.5%
I	θ_1	244.23	153.65	246.14	70.73	952.61
	θ_2	1587.49	1322.18	1080.72	321.76	4459.74
	θ_3	2121.27	1804.27	1353.79	552.48	5761.59
	β_1	0.00405	0.00221	0.00518	0.00013	0.01768
	β_2	0.00611	0.00368	0.00736	0.00006	0.02716
	β_3	0.00606	0.00366	0.0069	0.00007	0.0255
	τ_1	50.38	50.55	5.50	40.83	59.26
	τ_2	106.00	107.34	4.07	96.44	109.92
	α_1	1.263	1.232	0.335	0.716	2.022
	α_2	0.786	0.741	0.337	0.260	1.570
	α_3	0.797	0.768	0.284	0.336	1.512
	k_1	0.737	0.707	0.198	0.417	1.168
	k_2	0.093	0.063	0.096	0.013	0.339
	k_3	0.151	0.118	0.115	0.030	0.450
II	θ_1	271.77	209.24	205.09	79.88	856.36
	θ_2	456.94	211.64	556.71	78.73	2240.62
	θ_3	650.06	424.86	538.86	192.49	2201.55
	β_1	0.01021	0.00902	0.00578	0.00252	0.02462
	β_2	0.01179	0.00911	0.01005	0.00041	0.03658
	β_3	0.01031	0.00759	0.00963	0.00046	0.0368
	τ_1	49.20	48.45	5.53	40.55	59.20
	τ_2	101.05	101.12	4.73	92.18	108.27
	α_1	0.947	0.932	0.139	0.708	1.255
	α_2	0.867	0.859	0.297	0.320	1.452
	α_3	0.867	0.845	0.259	0.431	1.439
III	θ_1	221.80	171.96	153.60	85.79	700.27
	θ_2	184.82	143.24	130.11	89.92	598.77
	θ_3	397.48	331.83	226.51	231.97	1179.89
	β_1	0.01001	0.009	0.00567	0.00202	0.02358
	β_2	0.0097	0.0082	0.0069	0.00108	0.02642
	β_3	0.00638	0.00561	0.00412	0.00074	0.01603
	τ_1	50.00	49.85	5.53	40.79	59.13
	τ_2	102.16	102.68	4.55	92.58	108.37

$\theta_3 \sim Gamma(0.001, 0.001)$; $\beta_1 \sim Gamma(6, 1000)$; $\beta_2 \sim Gamma(1, 1000)$; $\beta_3 \sim Gamma(1, 100)$; $\alpha_1 \sim Gamma(1, 1)$; $\alpha_2 \sim Gamma(1, 1)$; $\alpha_3 \sim Gamma(1, 1)$; $\tau_1 \sim Uniform(2900, 2950)$ and $\tau_2 \sim Uniform(4340, 4360)$. The prior distributions for the parameters k_1, k_2 and k_3 are the exponential distribution with mean parameter 0.99 and truncated at 6 for all three parameters.

Figure 5 presents the empirical and estimated theoretical mean function m(t) assuming the models with no change-points and with one or two change-points. We can observe that the graphs of the function $m(t)$ estimated by the models without any change-points and with one and two change-points fit very well the data set. This result can be observed in the adjusted curves, which overlap the empirical curve.

m(t)

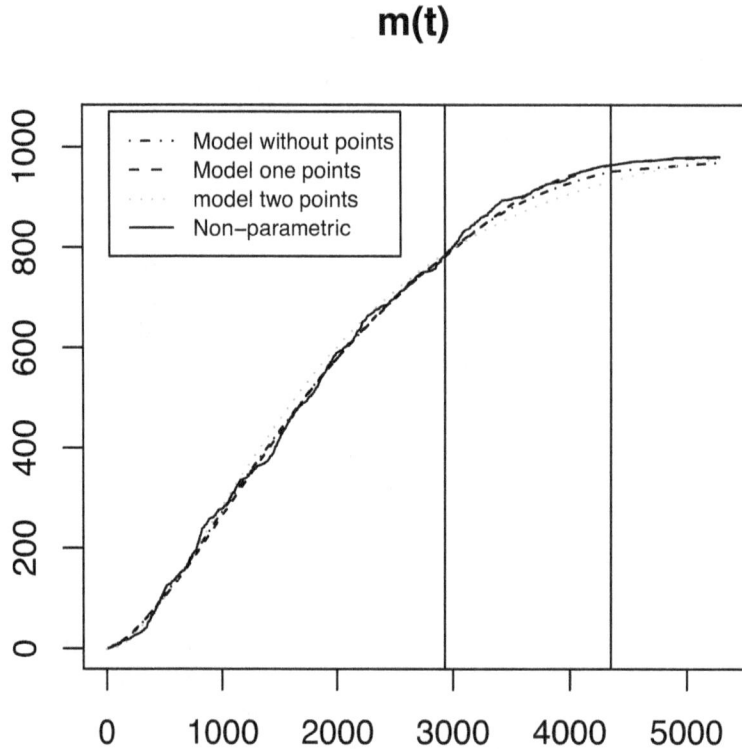

Figure 5: Ozone pollution from NE region of Mexico City. Graphs of the estimates of the mean function $m(t)$, empirical, and estimates by the models without any change-points and with one and two change-points. The vertical lines show the estimated change-points.

Table 3 presents the posterior summaries for the models without any change-point and with one and two change-points for Model I. Prior distributions for the parameters θ_i, β_i, α_i and τ_j, $i = 1, 23$, are presented in the appendix. The prior distributions for the parameters k_i, are those of the second proposed set of distributions, in which we have the prior exponential distribution with mean parameter 0.95 truncated at 3 for the model without change-point and with mean parameter 0.99 truncated at 6 for the models with one and two change-points.

We also use the Deviance Information Criterion (DIC) proposed by Spiegelhalter et al. (2002) to compare the models. This criterion is widely used in Bayesian model selection when samples of posterior distributions of parameters are obtained by MCMC simulation. The DIC selects the more complex models with two change-points, with DIC equal to -2413.32. The DIC for the models with no change-points and with one change-points are, respectively, 4863.14 and -2217.76.

Table 3:   Summary of estimates of the posterior distributions for the models without any change-point and with one and two change-points for Model I. Prior distributions for the parameters θ_i, β_i, α_i and τ_j, are presented in Section 3. The assumed prior distributions for the parameters k_i are the exponential distribution with mean parameter equal to 0.95 and truncated at 3 for the model without change-points and mean equal to 0.99 and truncated at 6 for the models with one and two change-points.

Model	parameter	mean	median	S.D.	2.5%	97.5%
without	θ	1035.04	1034.45	35.20	979.42	1093.69
change-points	β	0.0105081	0.0104119	0.0022471	0.0068893	0.0144018
	α	0.744	0.743	0.028	0.700	0.793
	k	2.898	2.925	0.093	2.710	2.995
one	θ_1	994.25	986.88	62.87	903.48	1105.67
change-point	θ_2	348.85	344.48	44.87	282.13	426.69
	β_1	0.0000100	0.0000097	0.0000025	0.0000067	0.0000146
	β_2	0.0000052	0.0000028	0.0000069	0.0000003	0.0000175
	τ	2923.45	2922.35	13.51	2902.91	2946.19
	α_1	1.501	1.502	0.038	1.437	1.561
	α_2	1.810	1.809	0.139	1.591	2.086
	k_1	1.000	0.999	0.045	0.929	1.075
	k_2	5.982	5.987	0.019	5.945	5.999
two	θ_1	1175.47	1166.73	107.61	991.77	1403.03
change-points	θ_2	3416.58	3215.40	1380.97	1331.79	6816.79
	θ_3	2354.08	1951.01	1448.58	601.19	5735.09
	β_1	0.00504	0.00471	0.00221	0.00167	0.0105
	β_2	0.00048	0.00038	0.00042	0.00007	0.00154
	β_3	0.0059	0.00381	0.00613	0.00015	0.023
	τ_1	2924.12	2923.43	13.40	2902.10	2946.76
	τ_2	4350.87	4350.88	4.99	4341.21	4359.29
	α_1	0.796	0.793	0.051	0.709	0.914
	α_2	1.110	1.104	0.111	0.890	1.314
	α_3	0.527	0.518	0.152	0.273	0.853
	k_1	2.298	2.290	0.213	1.917	2.730
	k_2	1.044	0.936	0.571	0.320	2.600
	k_3	0.186	0.139	0.168	0.026	0.597

6. Final Remarks

Considering the application with real data presented in Section 5, associated to air pollution in Mexico City, we observe that the intensity function of a NHPP could increase or decrease depending on the intervention measures adopted by the environmental authorities. In this case the use of NHPP processes with change-points to model the data could be an appropriate way to analyze the data set.

The use of a Bayesian approach based on MCMC methods for this kind of model could be very useful in the analysis of air pollution data. We also observed, from the obtained results of this paper, that sensitivity tests to different choices of prior distributions showed robust inferences, which guarantees the usefulness of the proposed methodology to analyze air pollution data.

The Bayesian sensitivity analysis was based on the effect on the posterior summaries using different choices of the hyperparameters for the prior distributions and also in the convergence of the MCMC algorithm. To illustrate the proposed methodology we also considered simulated data presented in Section 4.

The analysis was performed in two stages. Initially we worked with models without the presence of change-points; these models are of fundamental importance, because the information obtained from them serves to support more complex models which use change-points. The use of the information obtained in previous stages of the Bayesian analysis is important to achieve convergence of the MCMC chains.

The application of the generalized gamma model using Bayesian inference for NHPP with change-points proved to be an excellent tool to analyze the air pollution caused by ozone gas. It was observed that when applying more complex models, with the inclusion of change-points, the improvement in the fit tended to compensate for the increase in the number of parameters. This is also confirmed by the graphs of the MCMC chains of the parameters, the fitting of the curves and the DIC discrimination criterion.

Usually, it is observed that for air pollution data, the change-points occur after a period during which environmental action was taken by the government to reduce pollution levels. This is observed in the example of ozone air pollution in Mexico City presented in Section 5, where from 1990 to 1998 there was a reduction in the use of highly polluting vehicles and, around this period, the model detected a change-point. This was also observed in Achcar et al. (2010). It is noteworthy that, from 1999 to 2002, a series of measures was taken by the environmental authorities in Mexico, with the goal of reducing ozone levels; these measures included manufacturers being encouraged to produce cars with modern clean technology. The year of 2001 was a key date here. Matching this, around 2002 another turning point was detected.

Meteorological variables (temperature, wind speed, etc.) have a significant impact on daily ozone levels. This is beyond the goals of this paper and it is left as a goal of a future study.

Acknowledgments

This work was partially supported by grants from CNPq and FAPESP. We thank Guadalupe Tzintzun for providing the data set and Epifisma Lab (Unicamp). We thank two anonymous reviewers for the valuable comments and suggestions.

Appendix: Priors and MCMC sampling scheme

In the appendix it is presented the prior distributions and the conditional posterior distributions used in the MCMC sampling scheme.

Prior distributions for the model without change point

This work used two sets of a priori distributions for Model I; the first set (1) of a priori distributions are the distributions suggested by Achcar et al. (1998)

$$(i) \quad N' \sim \text{Poisson} \ \{\theta[1 - I_k(\beta T^\alpha)]\} \ ;$$
$$(ii) \quad \theta \sim \text{Gamma} \ (a, b); \text{ known a and b } ;$$
$$(iii) \quad \beta \sim \text{Gamma} \ (c, d); \text{ known c and d } ;$$

$$(iv) \quad \alpha \sim \pi_1(\alpha) \text{ where } \pi_1(\alpha) \propto \frac{1}{\alpha} \text{ for } \alpha(\alpha > 0);$$

$$(v) \quad k \sim \pi_2(k) \text{ where } \pi_2(k) \propto \frac{1}{k} \text{ for } k(k > 0).$$

In the second set (2), the prior distributions for α and k are modified as

$(iv) \quad \alpha \sim$ Gamma (e, f); known e and f ;

$(v) \quad k \sim$ truncated exponential distribution (a_n, g); known a_n and g.

$P(\lambda)$ denotes the Poisson distribution with parameter λ, a_n, g are the hyperparameters of the truncated exponential distribution, and a, b, c, d, e and f are known hyperparameters of the gamma distributions where $Gamma(a, b)$ denotes a gamma distribution with expected value $\frac{a}{b}$ and variance $\frac{a}{b^2}$. The choice of the first set of priors for the parameters of the model introduced by Achcar et al (1998) was based in the domain of the parametric space or using very non-informative priors (gamma priors for theta and β and improper priors for α and κ). In the second choice of the priors, the parameters e, f, g and a_n were chosen in order to have very non-informative priors (gamma priors for α, and for κ a truncates exponential such that the distribution is near uniform).

We assume independence of the prior distributions of parameters. The values of the hyperparameters are given in the applications. In the restricted Models II and III we use the same prior distributions for the free parameters.

Posterior distributions for the model without change point

The inference will be conducted based on information supplied by the posterior distribution of the parameters. Assuming independence of the prior distributions, the likelihood function for the Poisson processes, both for the homogeneous case and for the non-homogeneous case, are given as

$$L(\Theta \mid D_T) = \left\{ \prod_{i=1}^n \lambda(t_i) \exp\left(-\int_{t_{i-1}}^{t_i} \lambda(x) dx \right) \right\} \exp\left\{ -\int_{t_n}^T \lambda(u) du \right\},$$

that is,

$$L(\Theta \mid D_T) = \left\{ \prod_{i=1}^n \lambda(t_i) \right\} \exp\left\{ -\int_0^T \lambda(u) du \right\}.$$

Thus, we have

$$L(\Theta \mid D_T) = \left\{ \prod_{i=1}^n \lambda(t_i) \right\} \exp\left\{ -m(T) \right\}. \tag{15}$$

Initially, we assume the prior distribution for k proposed by Achcar et al.(1998). The posterior distribution is given by

$$P(N', \alpha, \beta, \theta / D_T) \propto \frac{\theta^{N'+n+a-1} \alpha^n \beta^{kn+c-1}}{N'! \{\Gamma(k)\}^n} \left\{ \prod_{i=1}^n t_i^{\alpha k-1} \right\} \{1 - I_k(\beta T^\alpha)\}^{N'}$$

$$\times e^{-(b+1)\theta - (d+\sum_{i=1}^n t_i^\alpha)\beta} \pi_1(\alpha) \pi_2(k). \tag{16}$$

Since the posterior distribution in (16) has no closed form, we resort to MCMC methods to simulate samples of the joint posterior distribution. The algorithm used to obtain posterior distribution samples in (16) is given as follows. For the subset of parameters whose full conditional posterior distributions is known, we sample directly from it, using the Gibbs Sampling algorithm and for: and for the subset of parameters in which the conditional densities are not known, the samples are simulated using the steps of the Metropolis-Hastings algorithm (see Metropolis et al., 1953; Hastings, 1970). Recommended references for the review of MCMC methods are given in Casella and Berger (2001). The required full conditional posterior distributions needed for the MCMC algorithms are given by,

(i) $N' \mid \theta, \alpha, \beta, k, D_T \sim Poisson[\theta(1 - I_k(\beta T^\alpha)];$

(ii) $\theta \mid N', \alpha, \beta, k, D_T \sim Gamma[a + n + N', b + 1];$

(iii) $\beta \mid N', \alpha, \theta, k, D_T \propto \beta^{kn+c-1} e^{-\beta[\sum_{i=1}^n x_i^\alpha + d]} \{1 - I_k(\beta T^\alpha)\}^{N'};$

(iv) $\alpha \mid N', \theta, \beta, k, D_T \propto \alpha^n \left\{ \prod_{i=1}^n t_i^{\alpha k - 1} \right\} e^{-\beta \sum_{i=1}^n t_i^\alpha} \{1 - I_k(\beta T^\alpha)\}^{N'} \pi_1(\alpha);$

(v) $k \mid \alpha, N', \theta, \beta, D_T \propto \frac{\beta^{kn}}{\{\Gamma(k)\}^n} \left\{ \prod_{i=1}^n t_i^{\alpha k - 1} \right\} \{1 - I_k(\beta T^\alpha)\}^{N'} \pi_2(k).$

The parameter θs have closed form distributions, so their posterior distributions are obtained through the Gibbs sampling method. For the other parameters we use the Metropolis-Hastings algorithm.

When we adopt the prior specification of the second set, proposed in Model I, we have that the posterior distribution is given by

$$P(N', \alpha, \beta, \theta / D_T) \quad \propto \quad \frac{\theta^{N'+n+a-1} \alpha^{n+e-1} \beta^{kn+c-1}}{N'! \{\Gamma(k)\}^n} \left\{ \prod_{i=1}^n t_i^{\alpha k - 1} \right\} \{1 - I_k(\beta T^\alpha)\}^{N'}$$

$$\times e^{-(b+1)\theta - (d + \sum_{i=1}^n t_i^\alpha)\beta - f\alpha - a_n k} I_{\{0,g\}}(k), \quad (17)$$

and the only changes in the full conditional posterior distributions occur in (iv) and (v), which are replaced by

(iv) $\alpha \mid N', \theta, \beta, k, D_T \propto \alpha^{n+e-1} \left\{ \prod_{i=1}^n t_i^{\alpha k - 1} \right\} e^{-\beta \sum_{i=1}^n t_i^\alpha - f\alpha} \{1 - I_k(\beta T^\alpha)\}^{N'};$

(v) $k \mid \alpha, N', \theta, \beta, D_T \propto \frac{\beta^{kn}}{\{\Gamma(k)\}^n} \left\{ \prod_{i=1}^n t_i^{\alpha k - 1} \right\} \{1 - I_k(\beta T^\alpha)\}^{N'} e^{-a_n k} I_{\{0,g\}}(k).$

Prior and posterior distributions for models with change-points

The prior distributions for the parameters are given as follows:

(i) $\tau_j \sim Uniform(f_j, g_j), j = 1, 2;$

(ii) $N_i' \sim Poisson[\theta_i(1 - I_{k_i}(\beta_i t_i^{\alpha_i}))], i = 1, 2, 3;$

(iii) $\theta_i \sim Gamma(a_i, b_i), i = 1, 2, 3, a_i, b_i$ known;

(iv) $\beta_i \sim Gamma(c_i, d_i), i = 1, 2, 3, c_i, d_i$ known;

(v) $\alpha_i \sim Gamma(e_i, h_i), i = 1, 2, 3, e_i, h_i$ known;

(vi) $k_i \sim$ truncated exponential $(m_i, u), i = 1, 2, 3$ and m_i, u known.

We also assume independence among the prior distributions.

Let N_i' be a latent variable that has Poisson distribution with parameters $P[\theta_i(1 - I_{k_i}(\beta_i \tau_j^{\alpha_i}))]$, and let $\boldsymbol{\Phi} = (\boldsymbol{w}, N_i')$. The inference is performed using information supplied by the posterior distribution of the parameters. Assuming independence among the prior distributions, in the presence of change-points the posterior and the prior distributions and the likelihood function are related as follows,

$$P(\mathbf{\Phi} \mid D_T) \propto L(D_T \mid \mathbf{\Phi})P(\mathbf{\Phi}), \qquad (18)$$

where $P(\mathbf{\Phi} \mid D_T)$ is the joint posterior distribution of $\mathbf{\Phi}$ conditional to data D_T; $P(\mathbf{\Phi})$ corresponds to the prior distributions of parameters and $L(D_T \mid \mathbf{\Phi})$ is the likelihood function. The full conditional posterior distributions needed for the Gibbs sampling algorithm are given by,

- $P(N_i') \sim P[\theta_i(1 - I_{k_i}(\beta_i \tau_j^{\alpha_i}))]$;

- $P(\theta_i) \sim Gamma(N_{\tau_j} + N_i' + a_i - 1, 1 + b_i)$;

- $P(\beta_i) \propto \beta_i^{K_i N_{\tau_j} + c_i - 1} e^{-\beta_i(\sum_{i=1}^{N_{\tau_j}} t_i^{\alpha_i} + d_i)} [1 - I_{k_i}(\beta_i \tau_j^{\alpha_i})]^{N_i'}$;

- $P(\alpha_i) \propto \alpha_i^{N_{\tau_j} + e_i - 1} \left(\prod_{i=1}^{N_{\tau_j}} t_i^{\alpha_i k_i - 1}\right) e^{-\beta_i \sum_{i=1}^{N_{\tau_j}} t_i^{\alpha_i} - h_i \alpha_i} [1 - I_{k_i}(\beta_i \tau_j^{\alpha_i})]^{N_i'}$;

- $P(k_i) \propto \left(\prod_{i=1}^{N_{\tau_j}} \frac{1}{\Gamma(k_i)}\right) \beta_i^{k_i N_{\tau_j}} \left(\prod_{i=1}^{N_{\tau_j}} t_i^{\alpha_i k_i - 1}\right) e^{-m_i k_i} I_{(0,u)} k_i [1 - I_{k_i}(\beta_i \tau_j^{\alpha_i})]^{N_i'}$;

- $P(\tau_j) \propto \left(\prod_{i=1}^{N_{\tau_j}} \frac{1}{\Gamma(k_i)}\right) \theta_i^{N_{\tau_j}} \beta_i^{k_i N_{\tau_j}} \alpha_i^{N_{\tau_j}} \left(\prod_{i=1}^{N_{\tau_j}} t_i^{\alpha_i k_i - 1}\right) e^{-\beta_i \sum_{i=1}^{N_{\tau_j}} t_i^{\alpha_i}}$

  $\times \left(\prod_{i=N_{\tau_j}+1}^{N_{\tau_j}} \frac{1}{\Gamma(k_i)}\right) \theta_i^{N_{\tau_j} - N_{\tau_{j-1}}} \beta_i^{k_i N_{\tau_j} - N_{\tau_{j-1}}} \alpha_i^{N_{\tau_j} - N_{\tau_{j-1}}} \left(\prod_{i=N_{\tau_j}+1}^{N_{\tau_j}} t_i^{\alpha_i k_i - 1}\right)$

  $\times e^{-\beta_i \sum_{i=N_{\tau_j}+1}^{N_{\tau_j}} t_i^{\alpha_i}} [1 - I_{k_i}(\beta_i \tau_j^{\alpha_i})]\}^{N_i'} [1 - I_{k_i}(\beta_i \tau_j^{\alpha_i}) + I_{k_i}(\beta_i \tau_{j-1}^{\alpha_i})]^{N_i'}$.

The parameters θ_i have distributions in closed form, so their posterior distributions are obtained using Gibbs Sampling, and for the other parameters α_i, β_i, k_i and τ_j the samples are obtained using the Metropolis-Hastings algorithm.

References

Achcar, J. A.; Dey, K.D. and Niverthi, M. – A Bayesian approach using nonhomogeneous Poisson process for software reliability models, In: A.S.Basu; S.K.Basu; S.Mukhopadhyay, (Org.). *Frontiers in Reliability*. ed.1 River Edge, NJ: World Scientific, v. 4, p. 1–18, 1998.

Achcar, J. A.; Rodrigues, E.R.; Paulino, C. D. and Soares, P. – Non-homogeneous Poisson models with a change-point: an application to ozone peaks in Mexico City, *Environmental and Ecological Statistics*, v. 17, p. 521–541, 2010.

Achcar, J. A.; Rodrigues, O.G. and Rodrigues, E.R. – The behaviour of a Metropolis-Hastings algorithm under different prior distributions: an application to ozone measurements in Mexico City, 3rd WSEAS International Conference on Urban Planning and Transportation (UPT'10), Corfu Island, Greece, p. 160–165, 2010.

Achcar, J. A.; Rodrigues, and Tzintzun, G. – Using non-homogeneous Poisson models with multiple change-points to estimate the number of ozone exceedances in Mexico City, *Environmetrics*, v. 22, p. 1–12, 2011.

Air Resource Board (ARB) – Review of air quality standard for ozone in California, Environmental Protection Agency, Staff Report, California, USA, 2005.

Álvarez, L.J.; Fernández-Bremauntz, A.A.; Rodrigues, E.R. and Tzintzun, G. – Maximum a posteriori estimation of the daily ozone peaks in Mexico City, *Journal of Agricultural, Biological and Environmental Statistics*, v. 10, n. 3, p. 276–290, 2005.

Bell, M.L.; Mcdermontt, A.; Zeger, S.L.; Samet, J.M. and Dominici, F. – Ozone and short-term mortality in 95 US urban communities, *Journal of the American Medical Society*, v. 292, p. 2372–2378, 2004.

Bell, M.L.; Peng, R. and Dominici, F. – The exposure-response curve for ozone and risk of mortality and the adequacy of current ozone regulations, *Environmental Health Perspectives*, v. 114, p. 532–536, 2005.

Bell, M.L.; Goldberg, R.; Rogrefe, C.; Kinney, P.L.; Knowlton, K.; Lynn, B.; Rosenthal, J.; Rosenzweweig, C. and Patz, J.A. – Climate change, ambient ozone, and health in 50 US cities, *Climate Change*, v. 82, p. 61–76, 2007.

Braga, A.L.F.; Zanobetti, A. and Schwartz, J. – The effect of weather on respiratory and cardiovascular deaths in 12 U.S. Cities, *Enviromental Health Perspectives*, v. 110, p. 859–863, 2002.

Brooks, Stephen P. and Andrew Gelman. – General methods for monitoring convergence of iterative simulations, *Journal of Computational and Graphical Statistics*, v. 7, p. 434–455, 1998.

Casella, G. and Berger, R. L. – *Statistical Inference*, Belmont, CA: Duxbury Press, second Edition, 2001.

Cox, D.R. and Lewis, P.A. – *Statistical Analysis of Series of Events*, London: Methuen Press, 1966.

Farhat, S.C.L.; Paulo, R.L.P.; Shimoda, L.M.; Conceição, G.M.S.; Lin, C.A.; Braga, A.L.F.; Warth, M.P.N. and Saldiva, P.H.N. – Effect of air pollution on pediatric respiratory emergency room visits and hospital admissions, *Brazilian Journal of Medical and Biological Research*, v. 38, p. 227–235, 2005.

Galizia, A. and Kinney, P.L. – Long-term residence in areas of high ozone: associations with respiratory health in a nationwide sample of nonsmoking young adults, *Environmental Health Perspectives*, v. 107, p. 675–679, 1999.

Gauderman, W.J.; Avol, E.; Gililand, F.; Vora, H.; Thomas, D.; Berhane, K.; McConnel, R.; Kuenzli, N.; Lurmman, F.; Rappaport, E.; Margolis, H.; Bates, D.; Peter, J. – The effects of air pollution on lung development from 10 to 18 years of age, *New England Journal of Medicine*, p. 1057–1067, 2004.

Gelman, A. and Rubin, D. R. A single series from the Gibbs sampler provides a false sense of security. Em *Bayesian Statistics* 4 (eds. J. M. Bernardo, J. O. Berger, A. P. Dawid and A. F. M. Smith), Oxford: Oxford University Press, p. 625–631, 1992.

Gelman, A. Inference and monitoring convergence. Em *Markov Chain Monte Carlo in Practice* (eds. W. R. Gilks, S. Richardson and D. J. Spiegelhalter), London: Chapman and Hall, p. 131–143, 1996.

Goel, A.L. and Okumoto, K. – An Analysis Of Recurrent Software Errors In A Real-Time Control System, In Proceedings of ACM Conference, Washington, D.C., USA, p. 496–501, 1979.

Goel, A.L. – A guidebook for software reliability assessment, Technical Report, RADC-TR-83-176, 1983.

Gouveia, N.; Freitas, C.U.; Martins, L.C. and Marcilio, I.O. – Hospitalizações por causas respiratórias associadas à contaminação atmosférica no município de São Paulo, Brasil, *Cadernos Saúde Pública*, v. 22, p. 2669–2677, 2006.

Hastings, W. K. – Monte Carlo sampling methods using Markov chains and their applications, *Biometrika*, v. 57, p. 97–109, 1970.

Javits J.S. - Statistical interdependencies in the ozone national ambient air quality standard, *Journal of the Air Pollution Control Association*, v. 30, p. 58–59, 1980.

Leemis, L. M. – Nonparametric estimation of the cumulative intensity function for a nonhomogeneous Poisson process, *Management Science*, v. 37, p. 886–900, 1991.

Loomis, D.P; Borja-Arbuto, V.H; Bangdiwalla, S.I. and Shy, C.M. – Ozone exposure and daily mortality in Mexico City: a time series analysis, Health Effects Institute Research Report, v. 75, p. 1–46, 1996.

Martins, L.C.; Latorre, M.R.D.O.; Cardoso, M.R.A.; Gonçalves, F.L.T.; Saldiva, P.H.N. and Braga, A.L.F. – Poluição atmosférica e atendimentos por pneumonia e gripe em São Paulo, Brasil, *Revista de Saúde Pública*, v. 36, p. 88–94, 2002.

Matthews D.E, Farewell VT. - On testing for a constant hazard against a change-point alternative, *Biometrics*, v. 38, p. 463–468, 1982.

Metropolis, N.; Rosenbluth, A.W.; Rosenbluth, M.N.; Teller, A.H.; Teller, E. –Equations of state calculations by fast computing machines, *Journal of Chemical Physics*, v. 21, p. 1087–1092, 1953.

Raftery A.E. and Akman, V.E. - Bayesian analysis of a Poisson process with a change-point, *Biometrika*, v. 73, p. 85–89, 1986.

Raftery A.E. - Are ozone exceedance rate decreasing?. Comment of the paper "Extreme Value Analysis of Environmental Time Series: An Application to Trend Detection in Ground-level Ozone" by R. L. Smith, *Statistical Sciences*, v. 4, p. 378–381, 1989.

Ruggeri F., Sivaganesan S. - On modelling change-points in nonhomogeneous Poisson processes, *Statistical Inference in Stochastic Models*, v. 8, p. 311–329, 2005.

Spiegelhalter, D.J.; Best, N.G.; Carlin, B.P. and Van der Linde, A. – Bayesian measures of model complexity and fit (with discussion), *Journal of the Royal Statistical Society*, B(64), p. 583–639, 2002.

Vicini, L.; Hotta, L. K. and Achcar, J. A. – Non-homogeneous Poisson processes applied to count data: a Bayesian approach considering different prior distributions, *Journal of Environmental Protection*, v. 3. p. 1336-1345, 2012.

Weinberg, J.; Brown, L. D.; Stroud, J. R. - Bayesian forecasting of an inhomogeneous Poisson process with applications to call center data, *Journal of the American Statistical Association*, v. 102, p. 1185-1198, 2007.

Wilson, R., Colome, S.D., Spengler, J.D. and Wilson, D.G. – *Health Effects in Fossil Fuel Burning: Assessment and Mitigation*, Cambridge, USA: Ballinger, 1980.

Yang T.E and Kuo L. - Bayesian binary segmentation procedure for a Poisson process with multiple change-points, *Journal of Computational and Graphical Statistics*, v. 10, p. 772–785, 2001.

Affiliation:

Lorena Vicini
Department of Statistics, UFSM
Avenida Roraima, 1000, CCNE
Santa Maria, RS, Brazil, 97105-900
E-mail: `vicini22@gmail.com`

Luiz Koodi Hotta
Department of Statistics, UNICAMP
Rua Sérgio Buarque de Holanda, 651
Campinas, SP, Brazil 13083-859
E-mail: `hotta@ime.unicamp.br`

Jorge Alberto Achcar
Department of Social Medicine
Medical School of Ribeirão Preto
Ribeirão Preto, SP, Brazil, 14049-900
E-mail: `achcar@fmrp.usp.br`

4

Optimal Deseasonalization for Monthly and Daily Geophysical Time Series

A. Ian McLeod and Hyukjun Gweon
Western University

Abstract

Deseasonalized geophysical time series are often used in time series models (Hipel and McLeod 1994). In this article an optimal method for selecting the deseasonalization transformation is suggested and an R package implementation (McLeod and Gweon 2012) is discussed. Our deseasonalization method may be used with the recently developed periodic autoregression model for daily river flow suggested by Tesfaye, Anderson, and Meerschaert (2011) and for the hierarchical Bayes modeling for multi-site daily temperature series discussed by Craigmile and Guttorp (2011).

Keywords: autoregressive model, harmonic regression, R, seasonality, time series modeling.

1. Introduction

Many geophysical time series are available on a monthly or daily basis and exhibit obvious seasonal features. These time series are often quite long. Lattice graphics capabilities available in R (R Development Core Team 2008) provide excellent multi-panel displays (McLeod, Yu, and Mahdi 2012). The built-in R function `stl()` provides a seasonal-trend decomposition that may be rendered in an attractive visual display as illustrated in McLeod *et al.* (2012) and is discussed in more detail by Cleveland (1993) for the famous (Wikipedia 2011) monthly Mauna Loa CO_2 time series. Long time series may also be visualized dynamically, like a movie, as illustrated in McLeod (2012b).

In this article our focus is on seasonal geophysical time series that are stationary after deseasonalizing by subtracting the seasonal mean and/or dividing by the seasonal standard deviation. [1] Suppose our time series consists of n successive monthly or daily values denoted

---

[1] Seasonal economic/financial time series are often more complicated due to non-stationarity and weekday/holiday artifacts.

by $z_t, t = 1, \ldots, n$ where t is the observation number. Then for modeling purposes it is often convenient to work with the deseasonalized version $w_t = (z_t - \mu_t)/\sigma_t$, where μ_t and σ_t are the seasonal mean and standard deviation. In the monthly case, often μ_t and σ_t are simply estimated by the monthly means and standard deviations. When the seasonal variances are constant, we may wish to use the detrended series, $w_t = z_t - \mu_t$.

Empirical (McLeod 1993) and theoretical (Ledolter and Abraham 1981) analyses have demonstrated that the principle of parsimony advocated for time series models by Box, Jenkins, and Reinsel (2005) is useful in selecting models that provide the best forecasts. In summary, this principle suggests that the time series model with the fewest number of parameters that adequately fits the data is preferred. Following this principle Hipel and McLeod (1994, §13.3.3) described a Fourier based approach for selecting the minimum number of Fourier components to use in the deseasonalizing transformation in the case of monthly hydrological time series. In the case of monthly time the seasonal frequency and its harmonic multiples, $12k/n$ are all Fourier frequencies so the problem reduces to an orthogonal regression that allows for efficient computation (Bloomfield 2004, Ch. 4). With daily time series it is natural to take the seasonal period to be $s = 365.25$ and so, in this case, the Fourier approach cannot be used but instead we may use an harmonic regression Craigmile and Guttorp (2011).

2. Harmonic Regression

In general, we may use harmonic regressions to estimate μ_t and σ_t. To estimate μ_t, we fit,

$$z_t = A_\mu^{(0)} + \sum_{k=1}^{F_\mu} \left(A_\mu^{(k)} \cos(2\pi kt/s) + B_\mu^{(k)} \sin(2\pi kt/s) \right) + u_t \qquad (1)$$

where $A_\mu^{(0)}$ is the overall mean, F_μ denotes the number of sinusoids used, $A_\mu^{(i)}, B_\mu^{(i)}, k = 1, \ldots, F_\mu$ are the sinusoid parameters, s is the seasonal period with $s = 12$ or $s = 365.25$ corresponding to the monthly and daily cases respectively, and u_t is the mean-zero error that is assumed to be stationary. It is well-known that in this case the least-squares estimates for the parameters $A_\mu^{(0)}, A_\mu^{(k)}, B_\mu^{(k)}, k = 1, \ldots, F_\mu$ are asymptotically fully efficient (Hannan 1970, §VII, Theorem 11). The estimated seasonal mean may be written,

$$\hat{\mu}_t = \hat{A}_\mu^{(0)} + \sum_{k=1}^{F_\mu} \left(\hat{A}_\mu^{(k)} \cos(2\pi kt/s) + \hat{B}_\mu^{(k)} \sin(2\pi kt/s) \right), \qquad (2)$$

where $\hat{A}_\mu^{(0)}, \hat{A}_\mu^{(k)}, \hat{B}_\mu^{(k)}$ are the least-squares estimates. Similarly the estimated seasonal variances,

$$\hat{\sigma}_t^2 = \hat{A}_\sigma^{(0)} + \sum_{k=1}^{F_\sigma} \left(\hat{A}_\sigma^{(k)} \cos(2\pi kt/s) + \hat{B}_\sigma^{(k)} \sin(2\pi kt/s) \right), \qquad (3)$$

where $\hat{A}_\sigma^{(0)}, \hat{A}_\sigma^{(k)}, \hat{B}_\sigma^{(k)}, k = 1, \ldots, F_\sigma$ are the least squares estimates in the regression,

$$\hat{u}_t^2 = A_\sigma^{(0)} + \sum_{k=1}^{F_\mu} \left(A_\sigma^{(k)} \cos(2\pi kt/s) + B_\sigma^{(k)} \sin(2\pi kt/s) \right) + v_t \qquad (4)$$

where $\hat{u}_t^2$ is the squared residual, $\hat{u}_t = (z_t - \hat{\mu}_t)^2$, and v_t is the mean-zero stationary error term. The deseasonalized time series, $w_t = (z_t - \hat{\mu}_t)/\hat{\sigma}_t$ may then be obtained. When $F_\mu = 0$, we set $\hat{\mu}_t$ equal to the sample mean of the original series z_t. Similarly when $F_\sigma = 0$, $\hat{\sigma}_t$ is set to the sample standard deviation of z_t.

As described in Hipel and McLeod (1994, §6.3 and §13.3) we may use the AIC (Akaike 1974) or BIC (Schwarz 1978) criterion to select F_μ and F_σ, the number of harmonics used. More generally we may use generalized AIC, defined as $\mathrm{GIC}_\alpha = -2\log L + \alpha k$, where L denotes the maximized value of the log-likelihood function and α is the tuning parameter with $\alpha = 2$ for the AIC and $\alpha = \log(n)$ for the BIC. Other choices for α have been discussed by (Taniguchi and Hirukawa 2012; Xu 2010; Xu and McLeod 2012).

For any fixed choices of F_μ and F_σ, the deseasonalized stationary time series, w_t, is assumed to be adequately modeled using an $AR(p)$. The R package FitAR (McLeod, Zhang, and Xu 2011) is used to automatically select p and determine the value of the GIC_α, denoted by $\mathrm{GIC}_\alpha(F_\mu, F_\sigma)$.[2] As F_σ changes, the scale changes for w_t and so it is necessary to adjust $\mathrm{GIC}_\alpha(F_\mu, F_\sigma)$ in order to be able to compare the effect of different choices of F_σ. This involves accounting for the transformation $w_t \longleftrightarrow z_t$ in the evaluation of the likelihood. The determinant of the logarithm of the Jacobian for the transformation $w_t \longleftrightarrow z_t$ is

$$\log J = -\sum_{i=1}^{n} \log \hat{\sigma}_t. \tag{5}$$

Hence the $\mathrm{GIC}_\alpha(F_\mu, F_\sigma)$ corresponding to a specific choice of F_μ and F_σ on the same scale as the original data z_t is given by

$$\mathrm{GIC}_\alpha^{(z)}(F_\mu, F_\sigma) = \mathrm{GIC}_\alpha(F_\mu, F_\sigma) - 2\log J, \tag{6}$$

where $\mathrm{GIC}_\alpha(F_\mu, F_\sigma)$ is computed using the transformed series.

For monthly time series we may enumerate $\mathrm{GIC}_\alpha^{(z)}(F_\mu, F_\sigma)$ for $F_\mu = 0, 1, \ldots, 6$ and $F_\sigma = 0, 1, \ldots, 6$ and select the optimal deseasonalization according to our chosen GIC_α-criterion. In this case $F_\mu = 6$ would correspond to simply using the monthly means while $F_\sigma = 6$ corresponds to using the monthly seasonal standard deviations. Note that $F_\mu, F_\sigma \leq 6$ to avoid aliasing (Bloomfield 2004; McLeod 2012a).

With daily time series often only a few harmonics are required for deseasonalization since the seasonal term is usually not too complicated. Hence we may choose a upper limits U_m and U_s and evaluate $\mathrm{GIC}_\alpha^{(z)}(F_\mu, F_\sigma)$ for $F_\mu = 0, \ldots, U_m$ and $F_\sigma = 0, \ldots, U_s$. In practice, $U_m = U_s = 6$ is often reasonable for many daily time series.

3. R Package

Our R package (McLeod and Gweon 2012) implements the methods discussed in §2. By default the BIC is used to select the optimum transformation although other criteria are available in the package as well.

Often it may be of interest to compare the optimum transformation with transformations that are close to it since if there is only a small difference, a simpler transformation may be more desirable. A good way to compare possible transformations is to use the relative plausibility.

---

[2] Occasionally, it may happen that $\hat{\sigma}_t^2 < 0$ defined in eqn. (3) is negative. This usually happens when F_σ or F_μ are too small or when the series needs a logarithmic or other type of transformation. When this happens we may simply set the value of the GIC_α to ∞ so this transformation will not be selected.

The relative plausibility of models, $i = 1, \ldots, I$ with generalized AIC, $\text{GIC}_\alpha(i), i = 1, \ldots, I$, is defined as

$$P_i = \exp\{-0.5(\text{GIC}_\alpha(i) - \text{GIC}_\alpha^\star)\}, \tag{7}$$

where $\text{GIC}_\alpha^\star = \min_i \text{GIC}_\alpha(i)$ (Akaike 1978; Hipel and McLeod 1994). This concept is similar to relative likelihood (Sprott 2000, §2.4). The output for the code snippets in §4 numerically illustrates the use of the relative plausibility.

All computations reported in §4 took only a few seconds. If necessary, with longer or more complicated seasonal time series or for Monte-Carlo applications, the enumeration to find the best GIC_α model could be vastly speeded up by using the R built-in package **parallel**.

4. Illustrative Examples

4.1. Monthly Saugeen River Flow

As an illustrative application, consider the mean monthly flow of Saugeen River at Walkerton in cumecs (m^3/sec) over the period from January 1915 to December 1979. There are $n = 744$ consecutive values. The ratio of maximum/minimum is about 56, so Tukey's rule-of-thumb (Tukey 1977, p. 397) suggests that a logarithmic transformation is in order and this was confirmed using a Box-Cox analysis (Hipel and McLeod 1994, §13.4.2). The lattice style boxplot in Figure 1 demonstrates that the log transformation makes the data distribution more symmetrical.

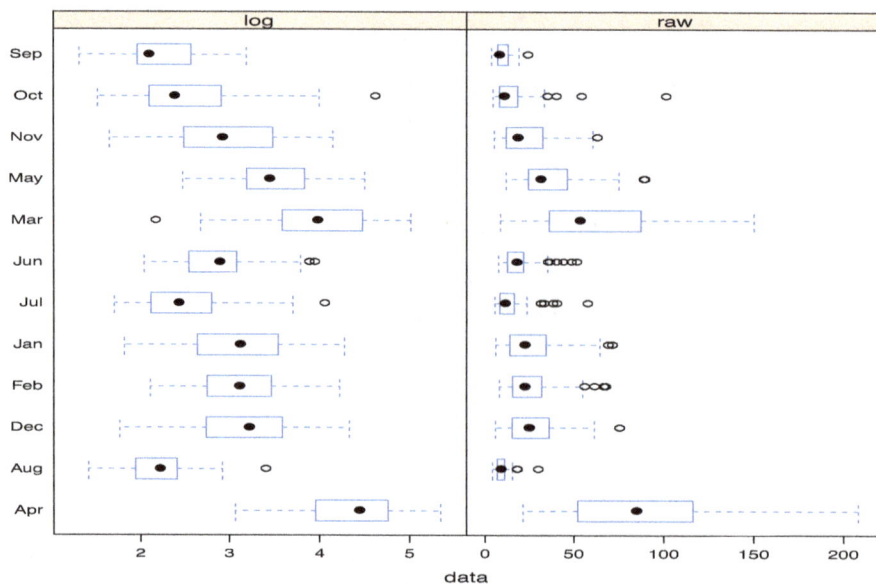

Figure 1: Comparing boxplots for original and log-transformed monthly flows for the Saugeen River

The lattice-style multipanel time series plot in Figure 2 shows the log series exhibits strong seasonality but no time trends or outliers.

Another useful plot for monthly time series obtained using the built-in R function `monthplot()` and is illustrated in Figure 3. This plot displays not only the seasonal pattern but the time

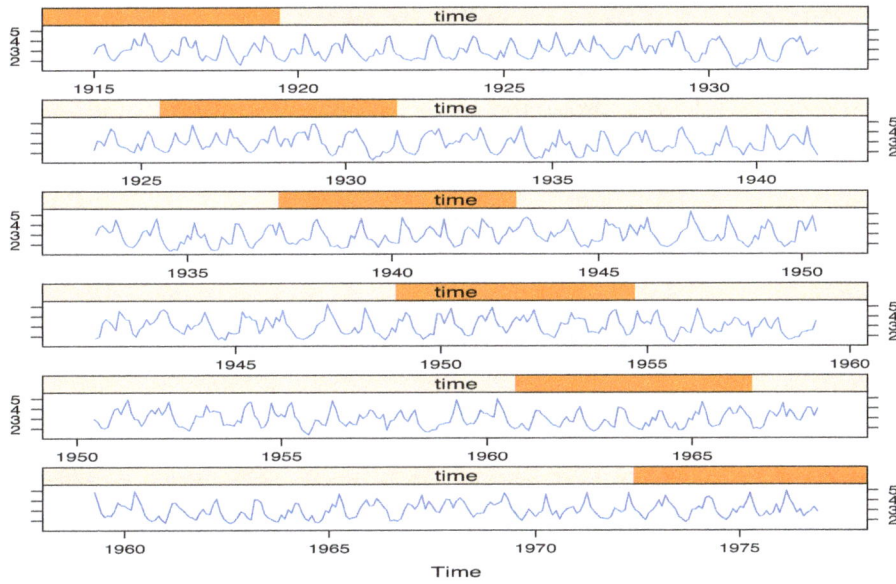

Figure 2: Lattice time series plot for monthly Saugeen flows.

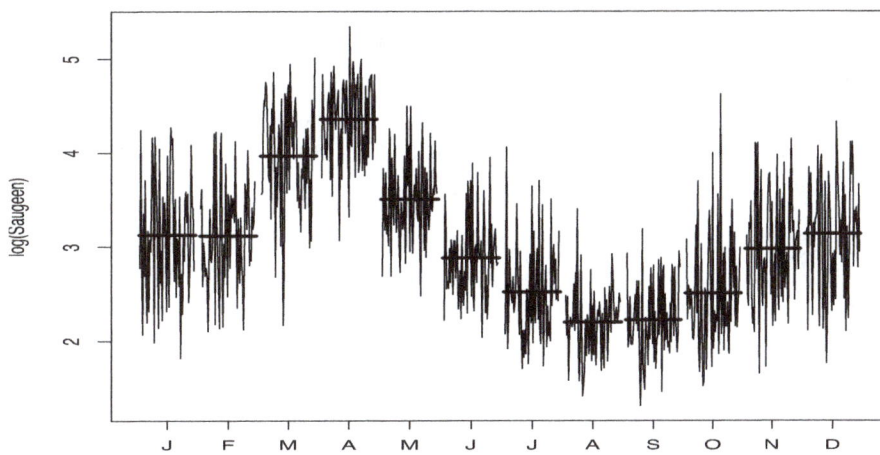

Figure 3: Monthplot for monthly Saugeen River flows.

series plot for each month separately. From this plot we don't see any evidence of trend-like changes occurring in the monthly subseries so the type of deseasonalizing transformation suggested in §2 is appropriate.

The output for the deseasonalization function, given in the code snippet below, shows that using the AIC results in $F_\mu = 5$ and $F_\sigma = 4$. This result agrees with that given by Hipel and McLeod (1994, §13.4.2) who also found that the optimal transformation was $F_\mu = 5$ and $F_\sigma = 4$ using an ARMA $(1, 1)$ model instead of selecting the best fitting AR. If the BIC criterion is used a more parsimonious deseasonalization with $F_\mu = 1$ and $F_\sigma = 1$ is obtained and only one other model has BIC-plausibility greater than 1%.

Code Snippets

The R code snippet below generates the lattice-style boxplot in Figure 1.

```
R >require("deseasonalize")
R >require("lattice")
R >n <- length(Saugeen)
R >i<-as.vector(cycle(Saugeen))
R >m<-month.abb[i]
R >Saugeen.df <- data.frame(z=c(Saugeen,log(Saugeen)), m=c(m,m),
+ which=rep(c("raw","log"), rep(n, 2)))
R >bwplot(m~z|which, data=Saugeen.df, scales=list(x=list(relation="free")),
+ xlab="data")
```

The following R command generates Figure 2,

```
xyplot(log(Saugeen),cut=TRUE)
```

The script below shows how monthly Saugeen river flow series is deseasonalized using the AIC criterion. The optimal transformation $F_\mu = 5$ and $F_\sigma = 4$ is indicated by the $*$ in the left column. The full output has been edited to show only the best 5 models as ranked by plausibility. This script took about 32 seconds. But when the BIC was used, the time was reduced to about 12 seconds. The difference in time reflects the fact that the BIC choose more parsimonious AR models than did the AIC.

```
R >out<-ds(log(Saugeen), ic="AIC")
R >summary(out)

  Fm Fs p        AIC Plausibility %
* 5  4 3 -1171.936          100.0
  6  4 3 -1171.065           64.7
  5  5 3 -1171.029           63.5
  5  3 3 -1170.261           43.3
  6  5 3 -1170.013           38.2
```

4.2. Daily Saugeen River Flow

A panel from a dynamic time series plot for a subseries of the mean daily flow Saugeen River at Walkerton, Jan 1, 1915 to Dec 31, 1979 is shown in Figure 4.[3] The series is comprised on 23,741 consecutive values.

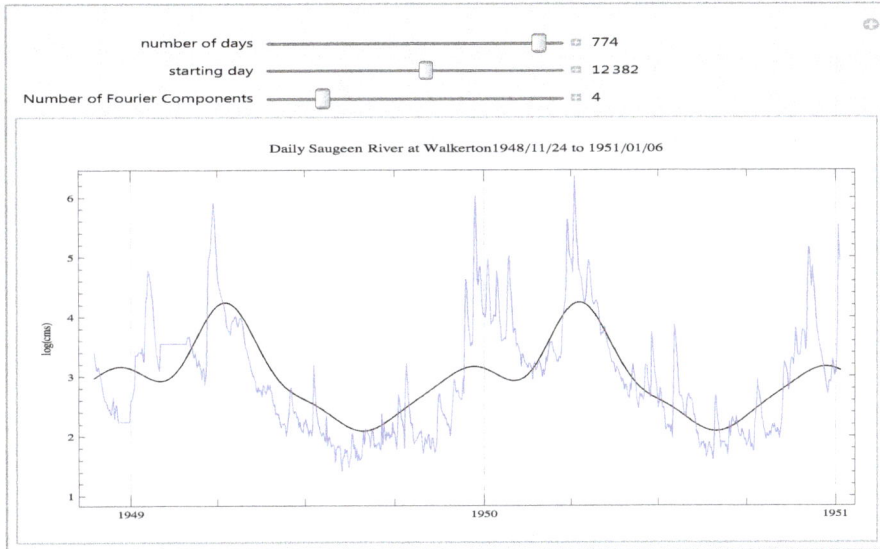

Figure 4: Dynamic time series plot of fitted harmonic regression to the daily Saugeen River flows.

For this series the BIC optimal deseasonalizing transformation was found to be with $F_\mu = 4, F_\sigma = 0$. All models with relative plausibility greater than 1% agreed with the choice $F_\sigma = 0$. The choice of the parameter F_μ may be explored visually using the dynamic time series plot illustrated in Figure 4.

As an additional check, the deseasonalized series was aggregated by month and Figure 5 shows resulting the boxplot. No seasonal variation is noticeable either the means or variances, so the deseasonalization appears to be effective.

Code Snippet

R script used to deseasonalize the Saugeen daily series. The output has been edited to show only the top five most plausible models. since the full output is rather length showing all combinations of $F_\mu, F_\sigma = 0, 1, \ldots, 6$. This script only took about 10 seconds which was less than for the monthly series. The reason for this is that the AR models selected were much simpler than in the monthly case.

```
R >out<-ds(log(SaugeenDay), Fm=6, Fs=6)
R >summary(out)

  Fm Fs p       BIC Plausibility (%)
*  4  0 6 -82621.80          100.0
```

[3] See the subdirectory /inst/doc located in the installation directory of our R package (McLeod and Gweon 2012) for instructions on how to view this plot dynamically on your computer.

```
5   0 6 -82617.37          10.9
3   0 6 -82615.34           4.0
6   0 6 -82613.19           1.3
```

The next script shows how the daily deseasonalized series is aggregated into months and the boxplot used to check for seasonality.

```
require("lubridate")
require("lattice")
w<-ds(log(SaugeenDay), Fm=4, Fs=0, searchQ=FALSE, standardizeQ=FALSE)$z
d<-rownames(SaugeenDay)
m<-month(d, label = TRUE, abbr = FALSE)
w.df <- data.frame(w=w, m=m)
bwplot(m ~ w, data=w.df, xlab="deseasonalized flow")
```

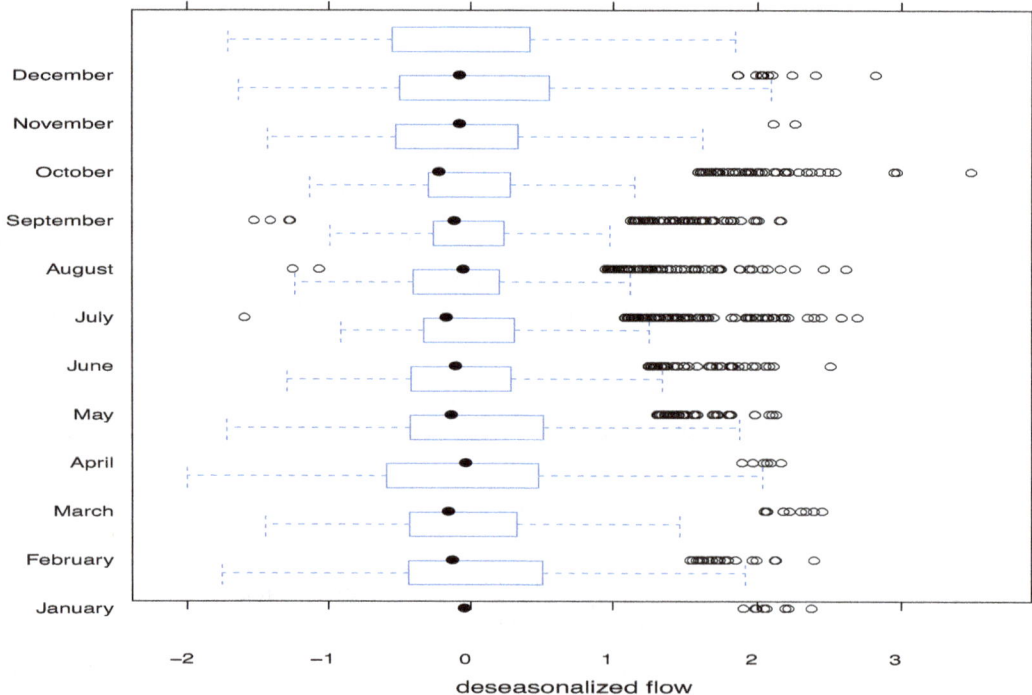

Figure 5: Boxplots of the detrended daily Saugeen River flows.

5. Concluding Remarks

The R function `stl()` provides another method for deseasonalizing monthly time series based on loess regression. This method is especially useful for time series that have a strong trend as well as a seasonal component but this method is less automatic and more complex than the method described in this article and it is only applicable to monthly series. The method we have described is preferable for time series models used for applications involving forecasting, simulation and intervention analysis as are described in Hipel and McLeod (1994) or in recent methods for time series modeling of daily series (Cressie and Wikle 2012; Craigmile and Guttorp 2011; Tesfaye *et al.* 2011).

Hipel and McLeod (1994); McLeod (1994) also discussed the application of periodic autoregression for modeling monthly geophysical time series. This type of correlation often occurs with river flow series when the spring runoff occurs either in March or April resulting in a reduced or even negative correlation between these two months whereas other months are positively correlated. The optimal selection using an AR(p) could be modified to use periodic autoregression. The R package **pear** (McLeod and Balcilar 2011) is available for fitting these models. But this approach is not likely to have any noticeable impact on the final deseasonalized series.

References

Akaike H (1974). "A New Look at the Statistical Model Identification." *IEEE Transactions on Automatic Control*, **19**(6), 716–723. [Accessed 14-Sept-2012], URL `http://dx.doi.org/10.1109/TAC.1974.1100705`.

Akaike H (1978). "A New Look at the Bayes Procedure." *Biometrika*, **65**(1), 53–59.

Bloomfield P (2004). *Fourier Analysis of Time Series: An Introduction.* 2nd edition. Wiley.

Box GEP, Jenkins GM, Reinsel GC (2005). *Time Series Analysis: Forecasting & Control (4th Edition).* Wiley, New York.

Cleveland WS (1993). *Visualizing Data.* Hobart Press.

Craigmile PF, Guttorp P (2011). "Space-time modelling of trends in temperature series." *Journal of Time Series Analysis*, **32**, 378–395. [Accessed 14-Sept-2012], URL `http://dx.doi.org/10.1111/j.1467-9892.2011.00733.x`.

Cressie N, Wikle CK (2012). *Statistics for Spatio-Temporal Data.* Wiley.

Hannan EJ (1970). *Multiple Time Series.* Wiley.

Hipel KW, McLeod AI (1994). *Time Series Modelling of Water Resources and Environmental Systems.* Elsevier, Amsterdam. [Accessed 14-Sept-2012], URL `http://www.stats.uwo.ca/faculty/aim/1994Book/default.htm`.

Ledolter J, Abraham B (1981). "Parsimony and Its Importance in Time Series Forecasting." *Technometrics*, **23**(4), 411–414. [Accessed 14-Sept-2012], URL `http://www.jstor.org/stable/1268232`.

McLeod AI (1993). "Parsimony, Model Adequacy and Periodic Correlation in Forecasting Time Series." *International Statistical Review*, **61**(3), 387–393. [Accessed 14-Sept-2012], URL http://www.jstor.org/stable/1403750.

McLeod AI (1994). "Diagnostic Checking Periodic Autoregression Models with Application." *Journal of Time Series Analysis*, **15**(6), 221–233. Addendum, Vol. 16, No. 2, p. 647, URL http://dx.doi.org/10.1111/j.1467-9892.1994.tb00186.x.

McLeod AI (2012a). *Aliasing in Time Series Analysis*. Wolfram Demonstrations Project. Accessed 03-April-2012, URL http://demonstrations.wolfram.com/AliasingInTimeSeriesAnalysis/.

McLeod AI (2012b). *Plotting a Long Time Series*. Wolfram Demonstrations Project. Accessed 03-April-2012, URL http://demonstrations.wolfram.com/PlottingALongTimeSeries/.

McLeod AI, Balcilar M (2011). *pear: Package for Periodic Autoregression Analysis*. R package version 1.2. Accessed 22-December-2011, URL http://CRAN.R-project.org/package=pear.

McLeod AI, Gweon H (2012). *deseasonalize: Optimal deseasonalization for geophysical time series using AR fitting*. R package version 1.31. Accessed 03-April-2012, URL http://CRAN.R-project.org/package=deseasonalize.

McLeod AI, Yu H, Mahdi E (2012). *Handbook in Statistics*, volume 30, chapter Time Series Analysis with R. Elsevier.

McLeod AI, Zhang Y, Xu C (2011). *FitAR: Subset AR Model Fitting*. R package version 1.92. Accessed 22-December-2011, URL http://CRAN.R-project.org/package=FitAR.

R Development Core Team (2008). *R: A Language and Environment for Statistical Computing*. R Foundation for Statistical Computing, Vienna, Austria. ISBN 3-900051-07-0, URL http://www.R-project.org/.

Schwarz G (1978). "Estimating the Dimension of a Model." *The Annals of Statistics*, **6**(2), 461–464. [Accessed 14-Sept-2012], URL http://projecteuclid.org/euclid.aos/1176344136.

Sprott DA (2000). *Statistical Inference in Science*. Springer.

Taniguchi M, Hirukawa J (2012). "Generalized Information Criterion." *Journal of Time Series Analysis*, **33**(2), 287–297. [Accessed 14-Sept-2012], URL http://dx.doi.org/10.1111/j.1467-9892.2011.00759.x.

Tesfaye YG, Anderson PL, Meerschaert MM (2011). "Asymptotic results for Fourier-PARMA time series." *Journal of Time Series Analysis*, **32**(2), 157–174. [Accessed 14-Sept-2012], URL http://dx.doi.org/10.1111/j.1467-9892.2010.00689.x.

Tukey JW (1977). *Exploratory Data Analysis*. Addison-Wesley.

Wikipedia (2011). "Keeling Curve." [Accessed 22-December-2011], URL http://en.wikipedia.org/wiki/Keeling_Curve.

Xu C (2010). *Model Selection with Information Criteria.* Ph.D. thesis, Western University. Electronic Thesis and Dissertation Repository. Paper 46. `http://ir.lib.uwo.ca/etd/46`.[Accessed 03-April-2012].

Xu C, McLeod AI (2012). "Further asymptotic properties of the generalized information criterion." *Electronic Journal of Statistics*, **6**, 656–663. [Accessed 14-Sept-2012], URL `http://projecteuclid.org/euclid.ejs/1334754009`.

Affiliation:

A. Ian McLeod
Department of Statistical and Actuarial Sciences
The University of Western Ontario
London, Ontario N6A 5B7 Canada
E-mail: `aimcleod@uwo.ca`

5

Space-time interpolation of daily air temperatures

M. Saez, M.A. Barceló
A. Tobias, D. Varga
GRECS, University of Girona

R. Ocaña-Riola
EASP, Granada

P. Juan, J. Mateu
University Jaume I, Castellón

Abstract

We propose a model to describe the mean function as well as the spatio-temporal covariance structure of 15 years of both maximum and minimum daily temperature data from 190 stations throughout the region of Catalonia (Spain), with daily data covering the period 1994-2008. Our aim is threefold: (a) estimation of the long-term trend of maximum and minimum temperatures; (b) assessing the spatial and temporal variability of temperatures, and (c) interpolation of the spatial temperatures at any given time.

Long-term trend, annual harmonics and winds were considered as explanatory variables of the mean function. The parameters associated with these variables were allowed to vary between stations and within each year. We controlled temporal autocorrelation by means of ARMA models. For the spatial covariance structure we used the Matérn family of covariance functions and a nugget term. Spatio-temporal models were built as Bayesian hierarchical models with two stages following the integrated nested place Laplace approximation (INLA) for Bayesian inference. For the final model estimation we used a two-stage approach, in which we first assumed the stations were spatially independent, and then we modeled the spatio-temporal covariance using the interim posterior from the residuals of the model in the first-stage as prior distributions of replications of a spatial process. We allowed all spatial parameters to also vary with time.

Keywords: Average temperature; Integrated nested Laplace approximation; Spatial variability; Spatio-temporal covariance; Two-stage Bayesian approach.

1. Introduction

The Intergovernmental Panel on Climate Change (IPCC), in its fourth and final evaluation report (IPCC, 2007a), pointed out several long-term changes in climate at global and regional scales. These changes have a higher probability of being associated with an anthropic activity.

Climate change refers to any significant change in measures of climate, such as temperature, precipitation and other weather patterns, that lasts for decades or longer. Consensus exists among scientists that the world's climate is changing, with more precipitation and weather extremes. Potential effects of this climate change are likely to include stronger and longer heat waves, more frequent heavy precipitation events, extreme weather events and increased air pollution (IWGCCH, 2010).

During the last years, efforts have focused on addressing how environmental changes can affect people's health. The more direct health effects of climate change can include injuries and illnesses from severe weather and heat exposure, increases in disease caused by allergies, respiratory problems, illnesses carried by insects or water and threats to the safety and availability of food and water supplies (IPCC, 2007b; IWGCCH, 2010).

Throughout the world, the prevalence of some diseases and other threats to human health depend largely on local climate, and particularly on local temperature. Climate-related disturbances in ecological systems can indirectly impact the incidence of serious infectious diseases, while extreme temperatures can lead directly to loss of life (IWGCCH, 2010). Many studies have shown that increases in average temperature may lead to more extreme heat waves during the summer, while producing less extreme cold spells during the winter. Higher temperatures, in combination with favorable rainfall patterns, could prolong disease transmission seasons in some locations where certain diseases already exist. In other locations, climate change will decrease transmission via reductions in rainfall or temperatures that are too high for transmission. Thus, temperature and humidity levels must be sufficient for certain disease-carrying vectors, such as ticks that carry Lyme disease, to thrive (National Research Council, 2001; IPCC, 2007b).

Additionally, temperature changes are expected to contribute to air quality and health problems. In fact, respiratory disorders may be exacerbated by warming-induced increases in the frequency of smog events and particulate air pollution (Schwartz and Randall, 2003). Sunlight and high temperatures, combined with other pollutants such as nitrogen oxides and volatile organic compounds, can cause ground-level ozone to increase. This increment can damage lung tissue, and is especially harmful for those with asthma and other chronic lung diseases. For other pollutants, the effects of climate change or weather are less studied and results vary by region. However, it seems clear that warm temperatures can increase air and water pollution, which in turn harm human health (McMichael et al., 2003; Schwartz and Randall, 2003).

Studying the local global climate change is a pressing challenge for public environmental and health agencies. The problem is broad and complex. However results are needed to respond to the challenges of global climate change (IWGCCH, 2010). The discovery of a significant trend towards an increase or decrease in the average values of a particular climatic element becomes a first symptom of climate change. The long-term evolution of the average temperature in a known time interval is considered a useful indicator that is easy to understand (Meteorological Service of Catalonia, 2010).

Motivated by these clear facts concerning the air temperature as an important climatic element, we analyze in this paper the spatio-temporal behavior of daily air temperatures in the Catalonian region of Spain. We propose a model to describe the mean function as well as the spatio-temporal covariance structure of 15 years of both maximum and minimum daily temperature data from 190 stations throughout the region of Catalonia (Spain, see Figure 1), with daily data covering the period 1994-2008. Our aim is threefold: (a) estimation of the long-term trend of maximum and minimum temperatures; (b) assessing the spatial and temporal variability of temperatures, and (c) interpolation of the spatial temperatures at any given time.

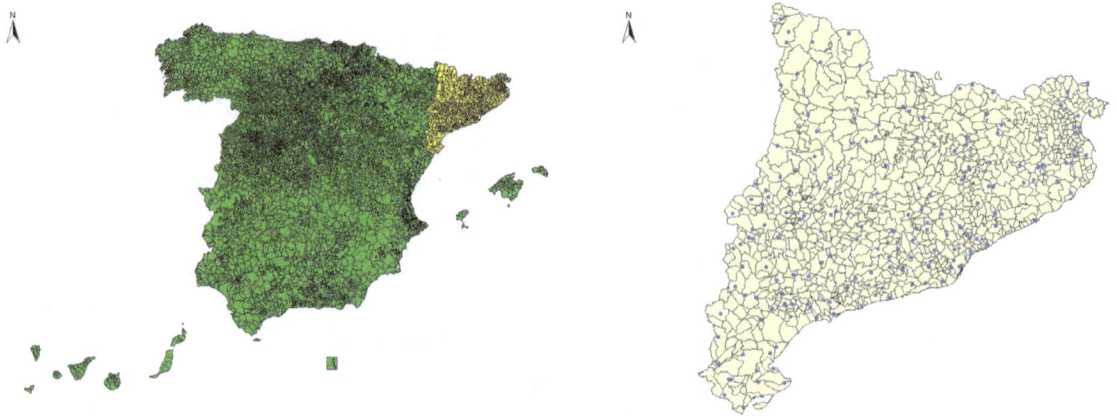

Figure 1: Situation of Catalonia region within Spain (*left*), and stations (in dotted points) with daily temperature data during 1994-2008 (*right*)

There are now many published research papers on interpolation of temperatures (Chessa and Delitala, 1997; Courault and Monestiez, 1999; Ninyerola *et al.*, 2000; Degaetano and Belcher, 2006; Stahl *et al.*, 2006, amongst many others). According to Stahl *et al.* (2006), schemes for spatial interpolation of air temperature vary in relation to three aspects: (1) the approach to adjusting for elevation, (2) the model used for characterizing the spatial variation of air temperature, and (3) the method of choosing prediction points. Most of them do not take into account the temporal dimension of the data. However, Luo *et al.* (1998) use a method that borrows information both across space and time. In fact, they could be considered a precedent of the related literature on spatio-temporal modeling of meteorological data (Haslett and Raftery, 1989; Huerta *et al.*, 2004; Lund *et al.*, 2006).

The statistical approach used in this paper is similar to the one used in Im *et al.* (2009), which, in turn, was similar to Li *et al.* (1999) to model particulate matter in Vancouver (Canada). Three important differences can be found in our work. First, unlike Im *et al.* (2009), and to get closer to reality, we allowed all parameters associated with the explanatory variables of the mean function of the temperature process to vary between stations and over time. Likewise, we allowed the ARMA models used to control autocorrelation to vary between stations, and the spatial parameters of our model to vary with time. Second, we followed a Bayesian approach and not a frequentist one, like in Im *et al.* (2009). Third, these authors estimate their model in three consecutive stages, using the residuals from the previous step as dependent variable. We however used a 2-stage approach. In the first stage, we assumed the stations were spatially independent and modeled the mean function of daily average temperature controlling for possible temporal autocorrelation. In the second stage, we modeled the spatio-temporal covariance using the interim posterior from the residuals of the model in the first stage as prior distributions of replications of a spatial process.

The plan of the paper is as follows. Section 2 presents the data set motivating this paper together with a description of the statistical strategy followed to model the data set in space and time. The Bayesian approach together with the INLA technique are discussed further in this Section. The results are commented in detail in Section 3. The paper ends with some final conclusions and discussion of the results.

2. Methods

2.1. Data set

Meteorological data, recorded daily for the period 1 January 1994 to 31 December 2008, from 190 stations throughout the region of Catalonia (Spain), were provided by the Weather Area (Meteorological Service of Catalonia) (see Figure 1). We had maximum and minimum temperatures, and wind (average wind speed and predominant direction) measurements for each station. The altitude and the spatial coordinates for each station were also considered.

2.2. Modeling strategy

The spatio-temporal process defining the daily temperature was specified following Im et $al.$ (2009)

$$
\begin{array}{rcl}
Y_{it} & = & X'_{it}\beta_{it} + \dfrac{\Phi_i(B)}{\Theta_i(B)}\varepsilon_{it} \\[2mm]
cov\left(\varepsilon_{it}, \varepsilon_{i't}\right) & = & \mathrm{M}\left(|i - i'|, r_t^2\sigma_t^2, \rho_t, \delta_t\right) + \left(1 - r_t^2\right)\sigma_t^2 I\,(i = i')
\end{array}
\tag{1}
$$

where Y_{it} denotes either the maximum or minimum temperature on day t (in our case, from 1 January 1994 to 31 December 2008) at station i with $i = 1, 2, \ldots, 190$. X_{it} contains the explanatory variables, while β is the unknown parameter vector associated to the explanatory variables. $\frac{\Phi_i(B)}{\Theta_i(B)}$ denotes the ARMA model, and ε_{it} stands for the innovations at each station on day t.

In particular, we included as explanatory variables of the mean function a long-term trend ($trend = 1, \ldots, 5479$, where 1 corresponded to 1 January 1994 and 5479 to 31 December 2008), annual harmonics $\cos\left(2\pi n\,trend/365\right)$ and $\sin\left(2\pi n\,trend/365\right)$, with $n = 1, 1.5, 2, 2.5, 3$ corresponding to periods $12, 8, 6, 4.8, 4$ months, respectively, and daily measurements of wind at each station, indeed a variable measuring the interaction between the average wind speed and the predominant direction -using a categorical variable capturing 8 sectors of the wind rose-. We considered more complicated forms than the linear approximation for the long-term trend, but these did not improve enough the goodness-of-fit of the models (perhaps because we already included in the model a 12-month period harmonic). In this sense, we preferred the parsimony of our approach to the complexity of the others.

The vector β of all parameters associated with the explanatory variables was allowed to vary between stations and with each year (this is the reason of the sub indexes in the parameters in (1)).

Meteorological conditions that may persist from one day to another can influence air temperature, leading to temporal correlation (Im *et al.*, 2009). We controlled this autocorrelation by means of ARMA models of the form

$$\frac{\Phi_i(B)}{\Theta_i(B)} = \frac{1 - \phi_{1i}B - \phi_{2i}B^2 - \dots - \phi_{pi}B^p}{1 - \upsilon_{1i}B - \upsilon_{2i}B^2 - \dots - \upsilon_{qi}B^q} \qquad (2)$$

where $\Phi_i(B)$ denotes the autoregressive (AR) component of order p; $\Theta_i(B)$ stands for the moving average (MA) component of order q; B is the backshift operator (i.e. $B^i X_t = X_{t-i}$), and finally ϕ and υ define the unknown AR and MA parameters, respectively. Note that again we allowed the parameters of the ARMA models to vary between stations (see the corresponding subindexes). In fact, we initially estimated ARMA(2,2) for all the stations, and according to the statistical significance of the parameters, the models were simplified.

For the spatial covariance structure we used the Matérn family of covariance functions and a nugget term, for a fixed t over each station i. In (1), M is the Matérn function (Stein, 1999), $\sigma_t^2 I$ denotes the *sill* (the total variance of the innovation process) at time t, $r_t^2\sigma_t^2$ is the variance of the spatially correlation portion of the process, $(1 - r_t^2)\sigma_t^2$ corresponds to the *nugget* (the variability for a given station), ρ_t is the *range* of the process (the size of the region where the process was significantly correlated), and finally δ_t is the smoothness degree of the process (we particularly tried $\delta_t = 1, 2, 3$, as we cover the most common and practical possibilities).

2.3. Estimation

The integrated nested Laplace approximation (INLA) for Bayesian inference

Spatio-temporal models were built as Bayesian hierarchical models with two stages (Schrödle and Held, 2010). The first stage was the observational model $\pi(y|x)$, where y denotes the vector of observations and x the vector of all Gaussian variables following a Gaussian Markov random field (GMRF). The second stage was given by the set of hyperparameters θ and their respective prior distribution $\pi(\theta)$.

Given data y for each component x_i of x, the marginal posterior density of the GMRF, $\pi(x_i|y)$, can be written as,

$$\pi(x_i|y) = \int_\theta \pi(x_i|\theta, y)\,\pi(\theta|y)\,d\theta \qquad (3)$$

Assuming that the precision matrix of the Gaussian field is sparse, the integrated nested Laplace approximation (INLA) for Bayesian inference (Rue *et al.*, 2009) build a nested approximation of (3). In particular, we can use the approximation by a finite sum

$$\tilde{\pi}(x_i|y) = \sum_k \tilde{\pi}(x_i|\theta_k, y)\,\tilde{\pi}(\theta_k|y)\,\Delta_k \qquad (4)$$

with $\tilde{\pi}\left(x_i\,|\theta,y\right)$ and $\tilde{\pi}\left(\theta\,|y\right)$ denoting approximations of $\pi\left(x_i\,|\theta,y\right)$ and $\pi\left(\theta\,|y\right)$, respectively. The posterior marginal $\pi\left(\theta\,|y\right)$ of the hyperparameters is approximated using a Laplace approximation (Tierney and Kadane, 1986)

$$\tilde{\pi}\left(\theta\,|y\right) \propto \frac{\pi\left(x,\theta,y\right)}{\tilde{\pi}_G\left(x\,|\theta,y\right)}\,|x = x^*\left(\theta\right) \qquad (5)$$

where the denominator $\tilde{\pi}_G\left(x\,|\theta,y\right)$ denotes the Gaussian approximation of $\pi\left(x\,|\theta,y\right)$ and $x^*\left(\theta\right)$ is the mode of the full conditional distribution $\pi\left(x\,|\theta,y\right)$ (Rue and Held, 2005).

According to Rue *et al.* (2009), it is sufficient to numerically explore this approximate posterior density using suitable support points θ_k, denoted as Δ_k.

In this paper, we defined these points in the h-dimensional space, using the strategy called central composite design (CCD). Centre points were augmented with a group of star points which allowed the estimation of the curvature of $\tilde{\pi}\left(\theta\,|y\right)$ (Rue *et al.*, 2009).

To approximate the first component of (4), we used a simplified Laplace approximation, less expensive from a computational point of view with only a slight loss of accuracy (Rue *et al.*, 2009; Martino and Rue, 2010).

Estimation strategy

Following the INLA approach, we specified the Matérn model for the latent Gaussian field (R-INLA, 2010) (subscripts were omitted for simplicity without loss of generality)

$$Corr\left(d\right) \propto \left(\kappa d\right)^\delta K_\delta\left(\kappa d\right) \qquad \alpha = \delta + d/2 \qquad (6)$$

where d denotes the spatial distance, K_δ defines the modified Bessel function, and δ sets the smoothness degree of the process. The range r was defined as $r = \sqrt{8}/\kappa$.

The hyperparameters of the model were the range and the precision parameter τ (the marginal variance of the latent field was $1/\tau$). Prior distributions were assigned to the log of the hyperparameters. In particular, Gamma distributions with parameters (i.e. shape and scale parameters) equal to 1 were considered. The smoothness parameter, δ, however, was considered a fixed parameter. As previously mentioned, we tried $\delta = 1, 2, 3$. In particular, δ was chosen to minimize the value of DIC (Spiegelhalter *et al.*, 2002)

$$DIC = D\left(\bar{\theta}\right) + 2p_D \qquad (7)$$

where $D\left(\bar{\theta}\right)$ denotes the deviance (defined as $D\left(\theta\right) = -2\log L\left(data\,|\theta\right)$, and where $L(.)$ denotes the (frequentist) likelihood function, and θ the parameters) evaluated at the posterior mean of the parameters, and p_D is the 'effective number of parameters' (which measures the complexity of the model), and is given by

$$p_D = E_{\theta|y}\left[D\right] - D\left(E_{\theta|y}\left[\theta\right]\right) = \bar{D} - D\left(\bar{\theta}\right) \tag{8}$$

with $\bar{D}$ denoting the posterior mean of standardized deviance.

The Matérn correlation function was built on a regular 200×200 lattice, covering the whole territory of Catalonia, corresponding to a cell of *1,207.31* m (height) $\times$ *1,226.65* m (base) (*1.481* km^2).

For the final model estimation we used a two-stage approach. In the first stage, we assumed the stations were spatially independent and we modeled the mean function of daily average temperatures controlling for possible temporal autocorrelation. In the second stage we modeled the spatio-temporal covariance, using the interim posterior from the residuals of the model (average within each cell) in the first stage as prior distributions of replications of a spatial process. For this second part we could realistically assume that the precision matrix of the Gaussian field was sparse (Rue and Held, 2005; Rue *et al.*, 2009). We allowed all spatial parameters to also vary with time.

3. Results

The analyses shown in this paper were carried out with the R freeware Statistical Package (version 2.11.1) (R Development Core Team, 2010) and the R-INLA package (R-INLA, 2010). The total number of measuring stations was 190 which recorded daily maximum and minimum temperatures during the study period 1994-2008.

The daily maximum temperature increased until 2003, decreasing thereafter (Figures 2 and 3). Note, however, that from 2003 onwards the variability of the daily maximum temperature increased dramatically, up to 15.26% each year, on average (Figure 3). This increase in the variability was also present in the daily minimum temperatures, although less important (8.76% on the average) and not so clear (Figures 2 and 3). Apparently, there was a decrease in daily minimum temperatures, much more important from 2003 onwards.

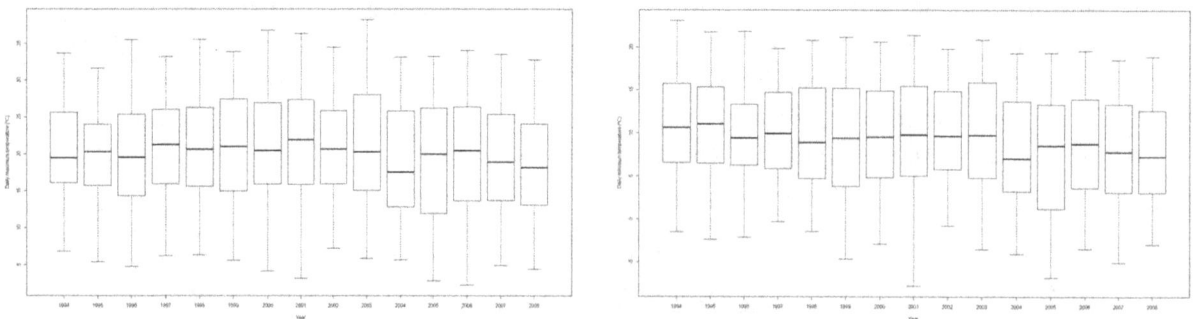

Figure 2: Boxplots by year of daily maximum (*left*) and minimum (*right*) temperatures in 190 stations in Catalonia for the period 1994-2008

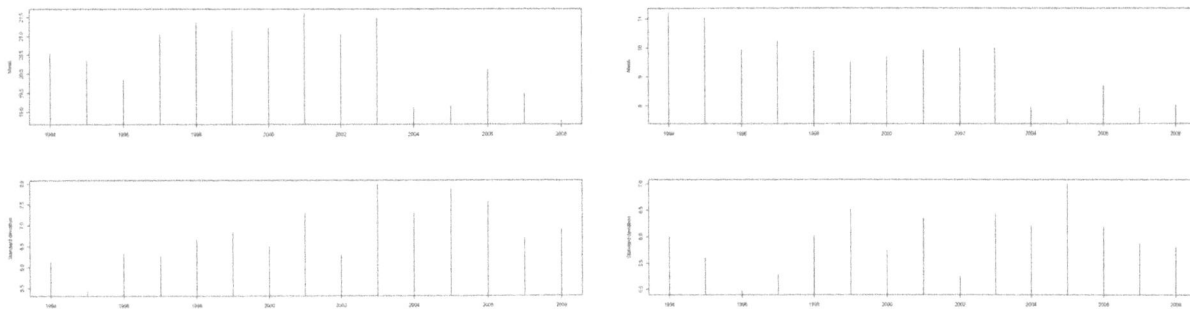

Figure 3: Annual averages and standard deviations of daily maximum (*left*) and minimum (*right*) temperatures in the 190 stations

With respect to the trend estimation in model (1), with daily maximum temperature as dependent variable, in 152 out of the 190 stations (80.0% of all the stations) the coefficient of the long-period trend was statistically significant at the 95% confidence level (i.e., the 95% credible interval did not contain the zero). In the case of daily minimum temperature, in 161 out of the 190 stations (84.7% of all the stations) the 95% credible interval of the coefficient of the long-period trend did not contain the zero.

All coefficients were 'averaged' using a linear mixed model, considering the station as grouping variable and the intercept as the random effect. In the model we allowed a within-group heteroscedasticity structure, with the standard errors of the coefficients as fixed variance weights. On 'average' the estimated coefficient for the long-period trend was equal to 2.109e-05, with a 95% credible interval (1.066e-05, 3.153e-05), for the maximum temperature, and a coefficient of 7.987e-05, with a 95% credible interval (1.446e-05,14.528e-05), for the minimum temperature.

These increases, however, were not homogeneous. First, the estimated daily variation in temperature depended on the latitude for both, maximum and minimum temperatures (Figure 4). In both cases the parametric coefficient of the relationship between (estimated) daily variation in temperature and latitude (in a generalized additive model with a Gaussian family and identity link) was statistically significant (p<0.001) but it was not the case for the smooth terms (p>0.1). The slope of the linear relationship was negative for maximum temperatures (-2.155e-05) but positive, clearly deeper (8.035e-05), for minimum temperatures (Figure 4). The estimated daily variation in temperature, however, did not depend on longitude or on altitude.

Estimated variation in daily temperature was not homogeneous in time either (Figure 5). Increases were only clear for the period 2004-2006 for maximum temperatures (median variation between 1994 and 2003, -0.036%, and 1.755% between 2004 and 2006), and for the period 2002-2006 when considering minimum temperatures (0.003% of median variation between 1994 and 2001, 1.349% between 2002 and 2006). Note also that the estimated variation was not homogeneous amongst months (Figure 6): (a) positive from April to September with

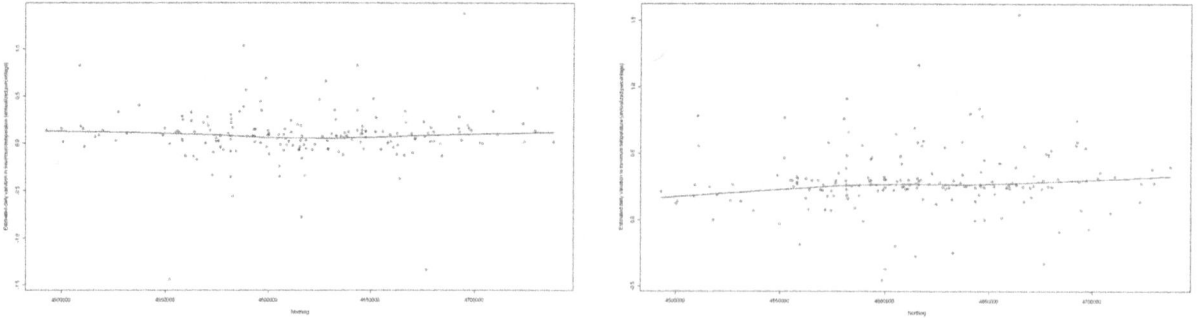

Figure 4: Estimated daily variation by latitude (annualized percentage) in maximum (*left*) and minimum (*right*) temperatures during the period 1994-2008

maximum temperatures (June, with a median increase of 12.74%, followed by May, with 11.89%, were the months with maximum increases and December, with -8.54%, and February, with -8.44%, were those months with the maximum decrease), and (b) positive from May to October with minimum temperatures (in this case it was July, with a median increase of 9.13%, and August, with 8.74%, the months with a maximum increase; and January, with -4.57%, February, with -4.11%, and December, with -3.98%, the months with the minimum variation).

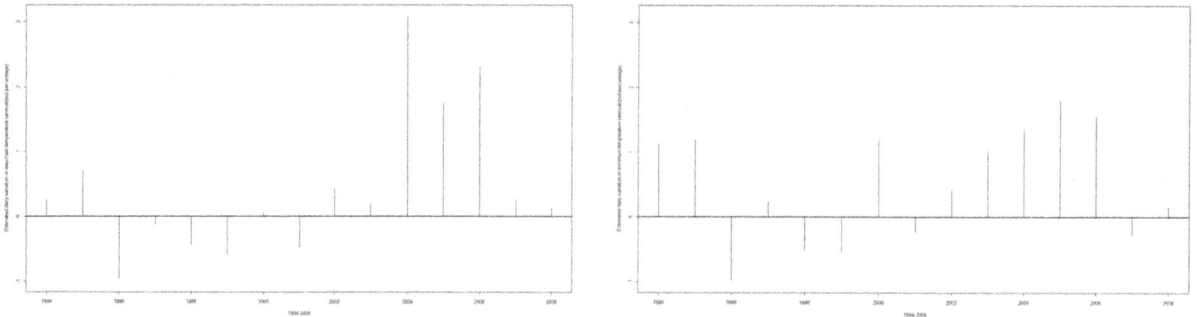

Figure 5: Estimated daily variation by year (annualized percentage) in maximum (*left*) and minimum (*right*) temperatures during the period 1994-2008

The coefficients associated with the harmonics were all also statistically significant. In addition, the temporal structure in form of ARMA models was the following. Note that we allowed the parameters of the ARMA models to vary between stations, and we initially estimated ARMA(2,2) for all the stations, and according to the statistical significance of the parameters, the models were simplified. We estimated an ARMA(2,1) model, i.e. $\frac{(1-\phi_1 B-\phi_2 B^2)}{(1-\upsilon_1 B)}$, with parameters ('averaged' as explained above) $\phi_1 = 0.68155$, $\phi_2 = -0.05743$, and $\upsilon_1 = -0.18330$, in 167 stations when considering maximum temperatures, and $\phi_1 = 0.32882$, $\phi_2 = 0.10523$,

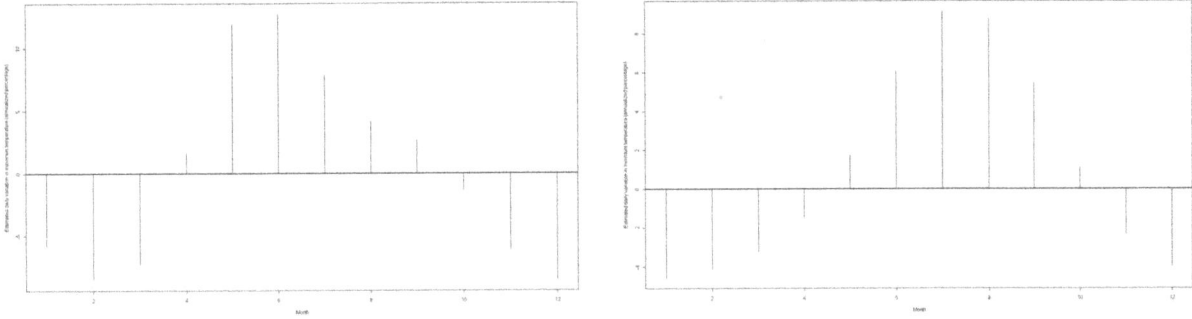

Figure 6: Estimated daily variation by month (annualized percentage) in maximum (*left*) and minimum (*right*) temperatures during the period 1994-2008

and $v_1 = 0.28255$ in 150 stations when using minimum temperatures. ARMA(1,1) models, i.e. $\frac{(1-\phi_1 B)}{(1-v_i B)}$, with parameters $\phi_1 = 0.48699$ and $v_1 = -0.01194$, were fitted in 19 stations for maximum temperatures, and in 14 stations for minimum temperatures with parameters $\phi_1 = 0.55320$ and $v_1 = 0.11454$. An AR(1) model, i.e. $(1 - \phi_{1i} B)$, with parameter $\phi_1 = 0.46448$ was fitted to 3 stations for maximum temperatures, and with $\phi_1 = 0.64679$ in 21 stations for minimum temperatures. Finally, a MA(1) model, i.e. $(1 - v_1 B)$, was fitted with parameter $v_1 = 0.41032$ in 1 station for maximum temperatures, and with $v_1 = 0.48629$ in 5 stations when considering minimum temperatures.

With respect to the estimation of the spatial bit of the model in (1), the value of the smoothness parameter δ that minimized DIC was equal to *2* for both maximum and minimum temperatures. The median estimated *sill* was (weighted average on time) *0.342* (first quartile *0.139*, third quartile *1.753*) for maximum temperatures, and it was *0.339* (first quartile *0.147*, third quartile *1.239*) for minimum temperatures. For both maximum and minimum temperatures, the median estimated *range* was *7,941* m (first quartile *7,908* m, third quartile *12,310* m). Finally, the median estimated *nugget* was practically negligible (*1.272e-12* for maximum temperatures, and *1.642e-12* for minimum temperatures).

Note that both the median and the variability of *sill* increased from 2004 onwards. In the case of maximum temperatures it increased from *0.2015* on average for the period 1994-2003 to *2.2579* for the period 2004-2008 (Figure 7), and from *0.1939* to *1.7047*, respectively, in the case of minimum temperatures (Figure 7).

In Figures 8 and 9 the septiles of the distribution of the interpolated standardized residuals (weighted average according to the temporal variation) are shown. In the case of maximum temperatures there was an increment of both the variability and the size of the residuals from the period 1994-2003 to 2004-2008. Note also that there was a movement of the upper septiles to the northeast and the south, from the west central and, in a lesser extent, the east central. The opposite behavior was found for minimum temperatures, i.e a slight decrease of both, the variability and the size of the residuals from the period 1994-2002 to 2003-2008, and a movement of the upper septiles to the north and the southeast, from the east and, in a lesser extent, the west central were found.

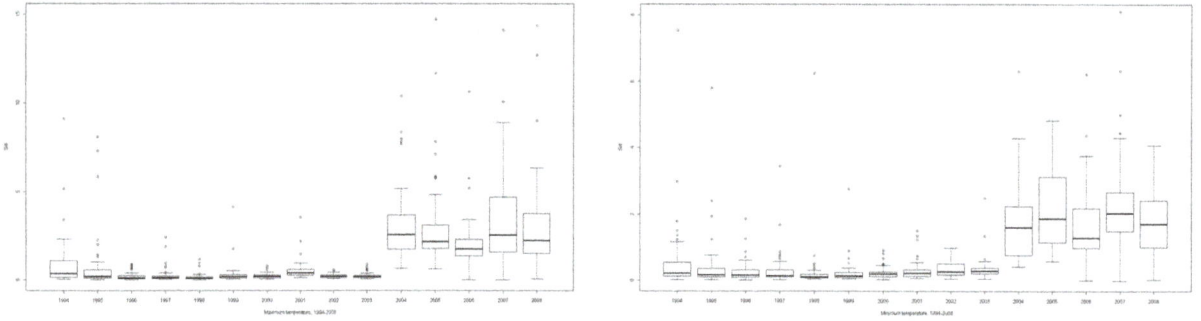

Figure 7: Estimated sill by year. Maximum (*left plot*) and minimum (*right plot*) temperatures

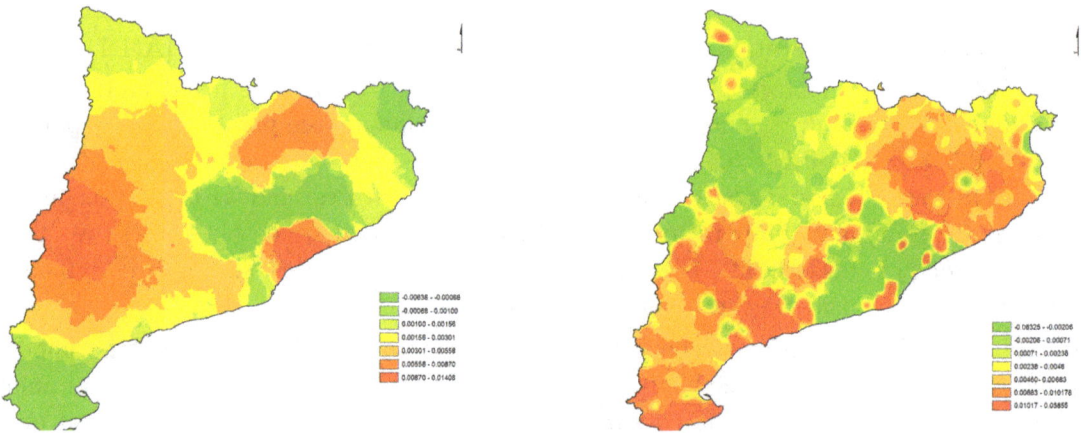

Figure 8: Standardized residuals for the maximum temperature for the period 1994-2003 (*left plot*), and for the period 2004-2008 (*right plot*)

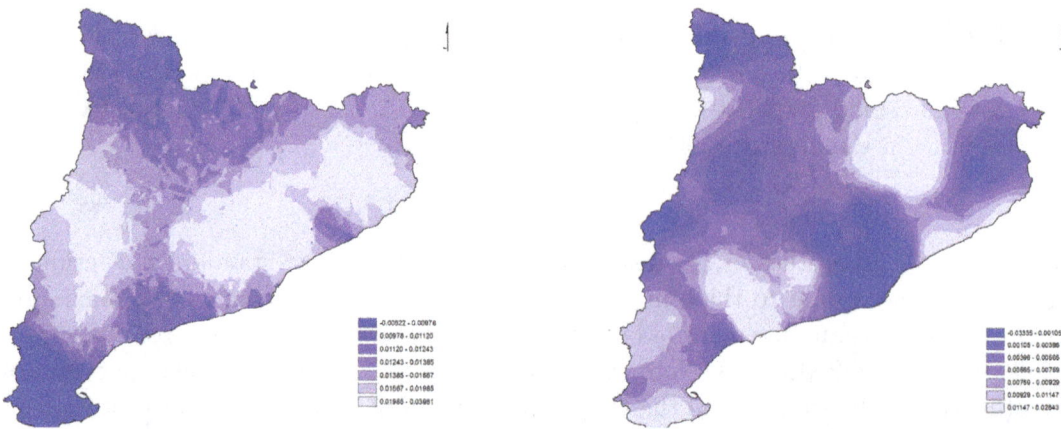

Figure 9: Standardized residuals for the minimum temperature for the period 1994-2002 (*left plot*), and for the period 2003-2008 (*right plot*)

4. Conclusions and discussion

Considering a model to describe the mean function as well as the spatio-temporal covariance structure of 15 years of daily maximum and minimum temperature data from 190 stations throughout Catalonia (Spain), with daily data covering the period 1994-2008, we have estimated a slight increase in daily maximum temperature (0.773% annualized on average) and a more significant increase in minimum temperature (2.960% annualized on average). These values correspond to a 15-year increase of 0.159^oC (95% credible interval 0.080^oC, 0.242^oC) in maximum temperature, and an increase of 0.332^oC (95% credible interval 0.008^oC, 0.635^oC) in minimum temperature.

For 60 years of daily temperature data from a subset of 16 stations in Catalonia (during the period 1950-2009), the Meteorological Service of Catalonia (2010) estimated a 10-year increase of 0.26^oC for the maximum temperature (in the range of $0.17^oC - 0.34^o$ C, depending on the station) and of 0.17^oC (range $0.09^oC - 0.23^oC$) for the minimum temperature. There were three differences with respect to our approach. First, the Meteorological Service of Catalonia (2010) estimated the long term trend for each station independently, without taking into account the spatial dimension of the data. Second, for the sub-period 1990-2009, they estimated a 10-year increase of 0.52^oC (range $0.07^oC - 0.90^oC$) for the maximum temperature, and 0.37^oC (range $0.05^oC - 0.64^oC$) for the minimum temperature. Finally, in their analysis they included the year 2009 (in the 16 stations they analyzed, 2009 was the third warmest year since 1950 - together with 1997 and, just after 2006 and 2003 - with an anomaly of 1.05^oC with respect to the climate average). In fact, they pointed out that positive trends obtained were slightly more pronounced, due to the warm character of 2009.

The increases we estimated were not homogeneous, both for time and space. Increases were only clear for the period 2004-2006 in maximum temperature, and for 2002-2006 in minimum temperature. In fact, considering only these periods, the increases we estimated practically matched those showed by the Meteorological Service of Catalonia (2010). Furthermore, in our case, the estimated variation was not homogeneous amongst months: positive from April to September in maximum temperature, and from May to October in minimum temperature. These variations were very similar to those obtained from the Meteorological Service of Catalonia (2010). We also found that the estimated daily variation in temperature depended on the latitude for both, the maximum and minimum temperature. The slope of the relationship was negative for the maximum temperature (-2.155e-05), and positive and clearly deeper (8.035e-05) for the minimum temperature. The Meteorological Service of Catalonia (2010) did not find spatial differences in the temporal variations of the stations they analyzed. It is very likely that the differences with respect to our work and the smaller number of stations they analyzed could explain this discrepancy.

5. Conflicts of Interest

There are no conflicts of interest for any of the authors. All authors disclose any financial and personal relationships with other people or organizations that could inappropriately influence and/or bias their work.

6. Acknowledgements

We would like to thank the Àrea de Climatologia, Servei Meteorològic de Catalunya for providing meteorological data. We are indebted to Natalia Salcedo who was (partially) responsible for the linguistic correction of our manuscript. We appreciate the comments of the attendees at the METMA V, International Workshop on Spatio-Temporal Modelling, in Santiago de Compostela, Spain, 30 June-2 July 2010, where a preliminary version of this work was presented. We also thank an anonymous referee that helped improve the manuscript. Work partially funded by grant MTM2010-14961 from the Spanish Ministry of Science and Education.

References

1. Chessa PA, Delitala AMS (1997). Objective analysis of daily extreme temperatures of Sardinia (Italy) using distance from the sea as independent variable. *International Journal of Climatology*, **17**, 1467-1485.

2. Courault D, Monestiez P (1999). Spatial interpolation of air temperature according to atmospheric circulation patterns in southeast France. *International Journal of Climatology*, **19**, 365-378.

3. Degaetano AT, Belcher BN (2006). Spatial interpolation of daily maximum and minimum air temperature based on meteorological model analyses and independent observations. *Journal of Applied Meteorology and Climatology*, **46**, 1981-1992.

4. Haslett J, Raftery AE (1989). Space-time modelling with long-memory dependence: assessing placecountry-regionIreland's wind power resource. *Applied Statistics*, **38**, 1-50.

5. Huerta G, Sansó B, Stroud JR (2004). A spatiotemporal model for Mexico city ozone levels. *Journal of the Royal Statistical Society, Series C*, **53**, 231-248.

6. Im HK, Rathouz PJ, Frederick JE (2009). Space-time modelling of 20 years of daily air temperature in the Chicago metropolitan region. *Environmetrics*, **20**, 494-511.

7. Intergovernmental Panel on Climate Change (IPCC) (2007a). *Climate Change 2007: The Physical Science Basis. Summary for Policymarkers*. World Health Organization-UNEP.

8. Intergovernmental Panel on Climate Change (IPCC) (2007b). *Climate Change 2007: Impacts, Adaptation and Vulnerability*. Cambridge: Cambridge University Press.

9. The Interagency Working Group on Climate Change and Health (IWGCCH) (2010). *A Human Health Perspective on Climate Change*. Washington, D.C.: Environmental Health Perspectives-National Institute of Environmental Health Sciences.

10. Li KH, Le ND, Sun L, Zidek JV (1999). Spatial-temporal models for ambient hourly PM_{10} in Vancouver. *Environmetrics*, **10**, 321-328.

11. Lund R, Shao Q, Basawa I (2006). Parsimonious periodic time series modelling. *Australian & New Zealand Journal of Statistics*, **48**, 33-47.

12. Luo Z, Wahba G, Johnson DR (1998). Spatio-temporal analysis of temperature using smoothing spline ANOVA. *Journal of Climate*, **11**, 19-28.

13. Martino S, Rue H (2010). Case studies in Bayesian computation using INLA. *Complex Data Modeling and Computationally Intensive Statistical Methods* (in press). URL: http://www.r-inla.org/papers.

14. McMichael AJ, Campbell-Lendrum DH, Corvalán CF, Ebi KL, Githeko A, Scheraga JD, Woodward A (2003). *Climate Change and Human Health: Risks and Responses.* World Health Organization. URL: http://www.who.int/globalchange/publications /cchhsummary/en/

15. Meteorological Service of Catalonia (2010). *Annual Bulletin of Climate Indicators. Year 2009.* Barcelona: Servei Meteorològic de Catalunya, Departament de Medi Ambient i Habitatge, Generalitat de Catalunya.

16. National Research Council (2001). *Climate Change Science: an Analysis of Some Key Questions.* Washington StateDC: National Academy Press.

17. Ninyerola M, Pons X, Roure JM (2000). A methodological approach of climatological modelling of air temperature and precipitation through GIS techniques. *International Journal of Climatology*, **20**, 1823-1841.

18. R Development Core Team (2010). *R: A language and environment for statistical computing.* R Foundation for Statistical Computing, Vienna, Austria. ISBN 3-900051-07-0. URL: http://www.r-project.org/.

19. R-INLA project (2010). URL: http://www.r-inla.org/home.

20. Rue H, Held L (2005). *Gaussian Markov Random Fields.* Boca Raton-London-New York-Singapore: Chapman and Hall/CRC.

21. Rue H, Martino S, Chopin N (2009). Approximate Bayesian inference for latent Gaussian models by using integrated nested Laplace approximations (with discussion). *Journal of the Royal Statistical Society, Series B*, **71**, 319-392.

22. Schrödle B, Held L (2010). Spatio-temporal disease mapping using INLA. *Environmetrics.* DOI: 10.1002/env.1065.

23. Schwartz P, Randall D (2003). *An Abrupt Climate Change Scenario and its Implications for United States National Security.*

24. Spiegelhalter DJ, Best NG, Carlin BP, Van der Linde A (2002). Bayesian Measures of Model Complexity and Fit (with Discussion). *Journal of the Royal Statistical Society, Series B*, **64**, 583-616.

25. Stahl K, Moore RD, Floyer JA, Asplin MG, McKendry IG (2006). Comparison of approaches for spatial interpolation of daily air temperature in a large region with com-

plex topography and highly variable station density. *Agricultural and Forest Meterology,* **139**, 224-236.

26. Stein ML (1999). *Statistical Interpolation of Spatial Data: Some Theory for Kriging.* New York: Springer.

27. Tierney L, Kadane JB (1986). Accurate approximations for posterior moments and marginal densities. *Journal of the American Statistical Association,* **81**, 82-86.

Affiliation:

Corresponding author: **Marc Saez**.

PhD, CStat, CSci Research Group on Statistics, Applied Economics and Health (GRECS) and CIBER of Epidemiology and Public Health (CIBERESP). University of Girona, Campus of Montilivi, E-17071 Girona, Spain. E-mail:marc.saez@udg.edu.

M.A. Barceló.

Research Group on Statistics, Applied Economics and Health (GRECS), and CIBER of Epidemiology and Public Health (CIBERESP). University of Girona, Spain.

A. Tobias.

Institute of Environmental Assessment and Water Research (IDAEA), Spanish Council for Scientific Research (CSIC), Barcelona, Spain, and Research Group on Statistics, Applied Economics and Health (GRECS), University of Girona, Spain.

D. Varga.

Research Group on Statistics, Applied Economics and Health (GRECS), and Geographic Information Technologies and Environmental Research Group University of Girona, Spain.

R. Ocaña-Riola.

Andalusian School of Public Health (EASP), Granada, Spain.

P. Juan.

Statistics and Operations Research, Department of Mathematics, University Jaume I, Castellón, Spain.

J. Mateu.

Statistics and Operations Research, Department of Mathematics, University Jaume I, Castellón, Spain.

6

Spatial Heterogeneity of the Nile Water Quality in Egypt

Amira El-Ayouti
Department of Statistics,
Faculty of Economics
and Political Science,
Cairo University.

Hala Abou-Ali
Department of Economics,
Faculty of Economics
and Political Science,
Cairo University.

Abstract

This paper aims to evaluate the water quality along the mainstream of the Nile in Egypt through modelling spatial distributions of water quality, using spatial statistical analysis. The study is based upon a sample frame of 78 sampling points collected in "February 2008" and located on the main waterway of the Nile and its delta (Rosetta and Damietta branches). Two water quality indices are calculated as general indicators of the overall water quality of the Nile, with special emphasis on drinking water quality. Exploratory spatial data analysis is carried out on the water quality indices, followed by plotting and modelling the experimental semi-variograms. Then, cross validation is executed in order to determine the best fitted models. Finally, surface maps are generated by performing spatial interpolation, using kriging technique. The generated surface maps of the two water quality indices show that water quality in Upper Egypt is excellent, in general, whereas water unfit for drinking is dominant in Middle and Lower Egypt. Therefore, intensive physical and chemical disinfection treatments are becoming pressing options for improving the quality of drinking water.

Keywords: The Nile, water quality, spatial modelling, semi-variogram, kriging.

1. Introduction

"Egypt gift of the Nile" said Herodot. The Nile constitutes the essential source of life in Egypt, it provides people with their fresh water needs. It is an essential factor of production and is vital for agriculture, transport, tourism and henceforth the socio-economic development of the country. However, the Nile has become, to a great extent, adversely affected by human activities. On the one hand, the population growth and the expansion of industrial, agricul-

tural, commercial and recreational activities that exploit natural resources, including water. On the other hand, industrial waste discharge, leakage of sewage by urban agglomeration and agricultural runoff contributes to the Nile contamination.

Therefore, the issue of Nile pollution should be on the top of the Government's environmental agenda. The protection of the aquatic environment requires regular water quality monitoring and effective pollution control in order to reduce the risk threatening the aquatic lives and people's health. Moreover, several quantitative research and statistical studies are needed to understand the intended problems, identify their limitations and, accordingly, propose realistic solutions.

To date, many studies try to make use of the spatial analysis methodology to model the spatial variations of the water quality indicators. The adopted method makes it possible to visualize the distribution of river water quality according to different land use and the interpolation of water quality at unsampled locations (Ouyang, Higman, Thompson, O'Toole, and Campbell 2001; Bordalo, Nilsumranchit, and Chalermwat 2001; French 2005; Sarangi, Madramootoo, and Enright 2006; Flipo, Jeannee, Poulin, Even, and Ledoux 2007; Rahman and Hossain 2008; Chang 2008)

Literature review reveals lack of studies that focus on water quality interpolation along the Nile through modelling the spatial variations of the water quality indicators. Therefore, this paper aims to fill this gap by mapping the water quality along the Nile, modelling the spatial variations and interpolating the water quality at unsampled locations. The software tool Geographic Information Systems (GIS) is used since the Nile has a geographical context. GIS is considered a powerful tool in managing water quality data, mapping and visualizing water quality spatial distributions (Elmahdi, Afify, and Abdin 2008; Hamad 2008).

Resorting to the interpolation techniques to illustrate the spatial variability is due to the difficulties of quantifying these variations at numerous locations. These difficulties are attributable to time constraints and impediments to access such locations. Thus, employing the interpolation technique help identify the locations with high concentration levels of pollutants. The generated surface maps, locating the spatial distribution of the water quality, will help decision makers adopt appropriate policies and undertake necessary measures not only to combat and prevent water pollution, but also to sustain this vital water resource, the Nile.

The remainder of the paper is organized to shed light on the study area and the available data. Section 3 provides the necessary back-ground of the applied methodology, taking into consideration the work-flow and the availability of reliable data as given in section 2. Results of the study are presented in section 4, and concluding discussion is given in section 5.

2. Data, study area and water quality indices

Although the Nile water quality is surely affected by the quality of water flowing from the upstream riparian countries, the current study only covers the Nile within the Egyptian territories. In Egypt, the Nile flows for a distance of about 1000 kilometers, starting from Aswan at $23^\circ 58' 24'' N$ and $32^\circ 53' 54'' E$ and ends into a large delta at where it flows into the Mediterranean Sea through Damietta branch at $31^\circ 31' 36'' N$ and $31^\circ 50' 41'' E$ and Rosetta branch at $31^\circ 27' 60'' N$ and $30^\circ 21' 53'' E$.

The Center of Environmental Monitoring and Studies of the Working Environment (CEM-SWE) monitors the water quality of the Nile through 78 monitoring sampling points located along the Nile main stream and its two branches. At each sampling location, monthly random samples of water are taken, and their physical and chemical properties are measured and recorded to assess the water quality. These points are taken as the sample frame of the study where 13 points are located in Greater Cairo, 45 points in Upper Egypt and 20 points in Lower Egypt (3 on Rosetta branch and 17 on Damietta branch). Each available sampling point has been identified by its longitude and latitude. Figure 2 portrays the map of Egypt and the distribution of the sampling locations along the Nile. At data collection stage, the latest available were for 2008. Thus, the study uses February 2008 as the target period since it lies within the dry season and is characterized by low water flow and no rainfall. These environmental conditions increase the level of pollution due to the absence of runoff water that aids in pollutants sequestration (Elmahdi *et al.* 2008).

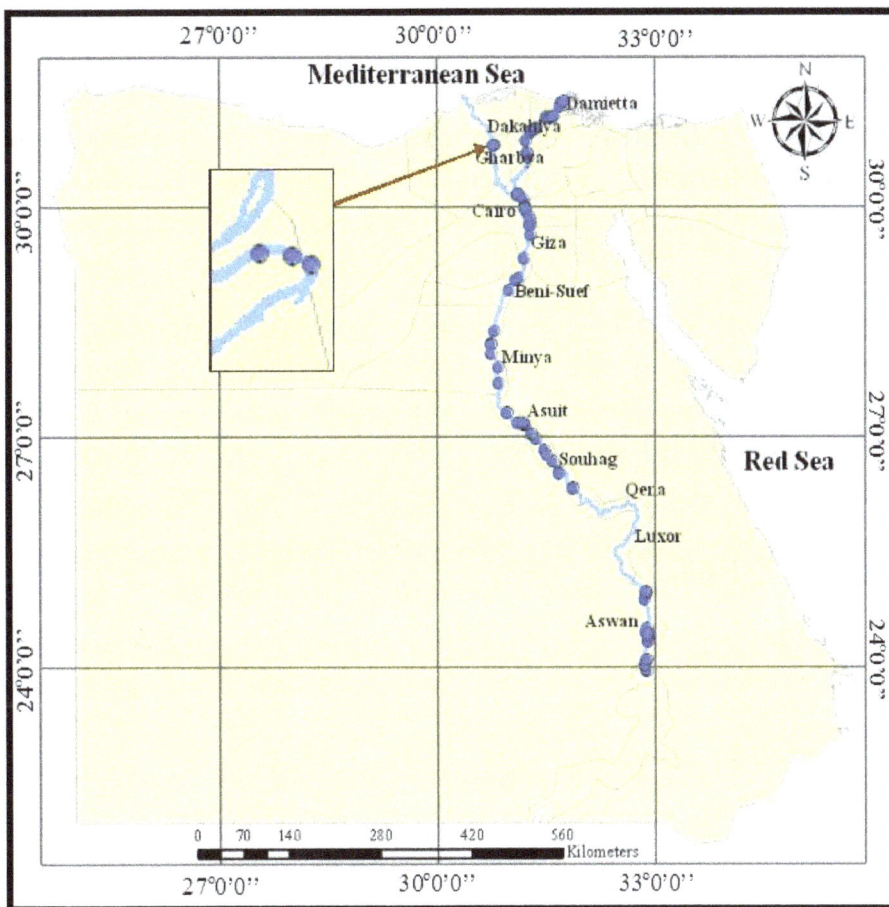

Figure 1: Map of the sampling points' distribution along the Nile

A water quality index (WQI) summarizes and streamlines complex water quality data into a single value that is easily conceivable. Once developed the WQI serves in examining trends, highlighting specific environmental conditions, planning water uses, as well as supporting decision making in assessing the worthiness of regulatory water quality program. In this research, the weighted water quality index (WWQI), proposed by Tiwari and Mishra (1985), is used to assess the overall water quality. Then, a drinking water quality index is developed to determine the suitability of water for this purpose. The WWQI determines the suitability of water to municipal uses, according to which, the water quality is rated excellent, good, poor, very poor and unfit when the value of the index lies between 0-25, 26-50, 51-75, 76-100 and >100, respectively. The WWQI is calculated as follows:

$$WWQI = \sum_{i=1}^{n} W_i Q_i, \tag{1}$$

where the subscript i denotes the ith water quality variable included in the index and Q_i is the variable quality rating calculated using the following formula:

$$Q_i = \frac{V_i - I_i}{S_i - I_i} * 100, \tag{2}$$

where V_i is the variable observed value at a given sampling site, I_i is the ideal value of the variable and S_i is the recommended water quality standard of the variable according to the Egyptian Law no 48 for 1982. Yet, W_i is the variable unit weight that is inversely proportional to the recommended water quality standard of the corresponding variable, i.e. $W_i = K/S_i$ such that K is a constant equal to $\frac{1}{\sum_{i=1}^{n} 1/S_i}$ and $\sum_{i=1}^{n} W_i = 1$. While, n is the number of water quality variables included in the index.

The drinking water quality index (DWQI) determines the suitability of water for drinking use. It is the average of the variables' quality ratings (Q_i's), calculated by Equation 2, taking into account that simple physical treatment and disinfection are used in the water treatment plants, i.e. using more rigorous standards than those used in the WWQI [1]. The DWQI is calculated as follows (Donia and Farag 2010):

$$DWQI = \frac{\sum_{i=1}^{n} Q_i}{number\ of\ available\ indicators}, \tag{3}$$

A DWQI value below 100 indicates that water is suitable for drinking after simple physical treatment and disinfection. But a DWQI value greater than 100 indicates that water is unfit for drinking use and therefore intensive physical and chemical treatment and disinfection are required in the water treatment plant.[2]

3. Methodology

Spatial analysis is the evaluation of data properties and relations, taking into account the spatial locality of the considered phenomenon. Spatial statistics assumes that the measured value of a variable z at a given sampling location x within a certain region D, $z(x)$, can be expressed as:

$$z(x) = m(x) + \varepsilon(x) + \varepsilon', \tag{4}$$

where $m(x)$ is the deterministic component of the variable at location x, $\varepsilon(x)$ is the spatially correlated error and ε' is the spatially independent error (Fortin and Dale 2005). The

---

[1]It should be noted that one of the drawbacks of these indices is the use of simple summation of subsidiary quality ratings. However, the WWQI or the DWQI are widely used in the literature due to their simplicity.

[2]For more details on these indices and their construction, consult Abou-Ali and El-Ayouti (2012).

methodology starts with exploratory spatial data analysis (ESDA) to explore the properties of the data, to test the underlying assumptions and to help in identifying the suitable model, followed by the spatial interpolation. The remainder of this section is organized in a manner that details these two steps.

3.1. Exploratory Spatial Data Analysis

This section summarizes the ESDA applied in order to ensure the suitability of data to implement spatial statistical analysis. This is a common practice that entails four steps. The first step inspects the normality of the data distribution. If data prove not normally distributed, hence a normality transformation is required. The second step is to detect outliers by the semi-variogram cloud since their presence may affect the analysis.

The third step involves trend investigation. Trend consists of two components: a fixed global trend and a random short range variation (random error). The first, if existent, should be removed, to fulfil the sationarity assumptions, then the random error is modelled (Hamad 2008). As a result, the trend analysis 3-D plot is used to identify the global trend. Also, Dowd (2003) introduced the global D-statistic to test for constant spatial mean as follows:

$$D_G = \frac{1}{n} \sum_{i,j=1}^{n} [z(x_i) - z(x_j)], \tag{5}$$

where n is the total number of sampling locations. The null hypothesis tested is the stationarity of the spatial mean. The test statistic is the standardized global D-statistic ($\check{D}_G = \frac{D_G}{\sqrt{\text{VAR}(D_G)}}$), which has asymptotically a standard normal distribution. The null hypothesis is not rejected with a confidence of $(100-\alpha)\%$ if the value of the test statistic is within the confidence interval, otherwise it is rejected, where α is the significance level.

Finally, the presence of spatial dependence, one of the most important properties of spatial data sets, is checked. The spatial dependence structure must be taken into account through modelling the spatial variations using semi-variogram models (Haining 1990). The degree of spatial dependence can be estimated using spatial autocorrelation coefficients such as Moran's I, which is defined as follows (Fortin and Dale 2005):

$$I = \frac{1}{w} \frac{\sum_{i=1}^{n}\sum_{j=1}^{n} w_{ij}[z(x_i) - \bar{z}][z(x_j) - \bar{z}]}{\frac{1}{n}\sum_{i=1}^{n}[z(x_i) - \bar{z}]^2}, \tag{6}$$

where $z(x_i)$ and $z(x_j)$ are the variable values at sampling location x_i and x_j, respectively and $\bar{z}$ is the sample mean of the variable. Yet, w_{ij} are the elements of the weight matrix such that $w_{ij} = \frac{1}{d_{ij}}$, where d_{ij} is the distance between x_i and x_j, and W is the sum of w_{ij}. Moran's I statistic ranges from -1 (negative correlation) to 1 (positive correlation). In this study, the geo-statistical analyst exploratory spatial data analysis toolbox of the Arc GIS 9.2 software is intensively used to apply the four main steps described above.

3.2. Spatial interpolation

The study applies the kriging interpolation technique, in which, the estimated value of the variable Z at a certain location x_o, $z^*(x_o)$, is a linear combination of the weighted average observations $z(x_i)$ at neighbouring locations x_i, $i = 1, 2, ..., m$ defined by: $z^*(x_o) = \sum_{i=1}^{m} w_i z(x_i)$. The weights w_i are based on the distance and the structure of spatial dependence between observations (Hamad 2008). In fact, kriging technique has many advantages, of which: 1) it provides the best linear unbiased estimator for $Z(x_o)$; 2) it incorporates the spatial variability to enhance the prediction efficiency; and 3) it is accompanied with a measure of precision. Therefore, kriging has been considered the most suitable spatial interpolation technique to be used in this study.

Kriging is divided into two tasks: 1) modelling the spatial structure of the data, using semi-variograms and 2) interpolating the value of a certain variable at an unobserved location, where the weights assigned to the observations are determined on the basis of the fitted semi-variogram model. The semi-variogram is a function describing the spatial variability between the variable values at different locations within the study area. It is defined as $\gamma(h) = 1/2\,Var[Z(x) - Z(x + h)]$ and estimated using the following semi-variance function:

$$\widehat{\gamma}(h) = \frac{1}{2N(h)} \sum_{i=1}^{N(h)} \left[z(x_i) - z(x_i + h)\right]^2 \tag{7}$$

where $z(x_i)$ is the value of the variable Z at the sampling location x_i, and $N(h)$ is the number of pairs of sampling locations located at distance h from one another (Fortin and Dale 2005). Yet, h is the spatial lag size used to reduce the larger number of possible combinations. Trial and error approach is usually considered in the selection of the lag size and the number of lags (Johnston, Hoef, Krivoruchko, and Lucas 2001; Sarangi et al. 2006).

After computing the "experimental semi-variogram" from the sample data, a "theoretical semi-variogram" is modelled to fit this experimental semi-variogram, through estimating three parameters the sill, the range and the nugget. Since this study is limited to the longitudinal profile of the Nile, the semi-variograms are estimated and fitted in one direction only (i.e. isotropy is assumed). The best semi-variogram model and its parameters are evaluated using cross-validation method, by ignoring each data point, one at a time, and kriging the associated data value. Then, the differences between the interpolated and the observed values are summarized using cross-validation statistics, namely, the mean error (ME), the root mean squared error (RMSE), the mean standardized error (MStE) and the root mean squared standardized error (RMSStE) (Johnston et al. 2001).

In the kriging family, ordinary kriging is the most commonly type used. This study uses ordinary kriging, which assumes the model $Z(x) = \mu + \varepsilon(x)$, where $\varepsilon(x) \sim (0, \Sigma)$, μ is an unknown constant mean and $Z(x)$ is weakly stationary. The optimal weights w_i, $i = 1, 2, \ldots, m$, that will yield the best linear unbiased estimate for the value of the variable of interest at one or more unsampled locations, can be obtained by minimizing the kriging variance $\sigma_E^2 = V[Z^*(x_o) - Z(x_o)]$, subject to the unbiasedness condition $\sum_{i=1}^{m} w_i = 1$. The estimated weights are then substituted in $z^*(x_o) = \sum_{i=1}^{m} w_i z(x_i)$ in order to obtain the interpolated value $z^*(x_o)$.

In brief, after computing the experimental semi-variogram, different semi-variogram models are fitted to the experimental semi-variogram to select the best fit model and its parameters, using cross-validation techniques. Finally, interpolating the variable values at unsampled locations and generating surfaces illustrating the spatial distributions of the variable under study can be performed, using kriging interpolation technique. Also, ArcGIS geo-statistical toolbox has been used for spatial interpolation, using the ordinary kriging module, for the two water quality indices: WWQI and DWQI, taken one by one.

4. Results

4.1. Exploratory Spatial Data Analysis

Following the four stage of ESDA previously described. Starting with testing the normality assumption through the inspection of the normal Q-Q plots indicate that the two WQIs are not normally distributed. Applying a log transformation to the WQIs ensures the normality of the variables.

Second, the 3-D trend analysis plots indicate that the WWQI and the DWQI decrease from Upper Egypt to Middle Egypt and Lower Egypt. However, the results of the D-statistic test support the stationarity of the WWQI (D = 1.83, p-value = 0.27) and the DWQI (D = 1.9, p-value = 0.23). Accordingly, there is no need to remove the first order polynomial trend. Thus, it can be concluded that the WQIs are stationary and it is better to use ordinary kriging assuming constant trend to interpolate their quality levels along the Nile.

Third, scrutinizing the semi-variogram clouds of the WQIs reveals the presence of outliers. However, these outliers have no significant effect on the degree of spatial dependence measured by Moran's I with a value of about 0.5 and a p-value < 0.001. Hence, no need to delete them from further analysis. Moran's I statistics indicate the significant presence of positive moderate spatial autocorrelation i.e. near locations are more related than distant locations.

4.2. Spatial interpolation

By trial and error, the lag size is set to 20 km and the number of lags is chosen in a manner that the distance of significant autocorrelation becomes clearly visible. Figure 2 illustrates both the experimental and the theoretical best fitted semi-variogram model for each variable, assuming constant trend. Generally, each semi-variogram starts low at closer distances and elevates as the distance widen. The spatial analysis results indicate that the best theoretical models (based on the root mean square errors) which fit the experimental semi-variograms are the Gaussian models for both WQIs given by:

$$\gamma(h) = \begin{cases} C_o + C_1 \left[1 - \exp\left(-3\frac{h^2}{a^2}\right)\right] & for\ 0 < h < a \\ C_o + C_1 & for\ h \geq a \end{cases}, \tag{8}$$

where C_o, C_1, a indicate the nugget, the sill and the range, respectively and h is the lag size.

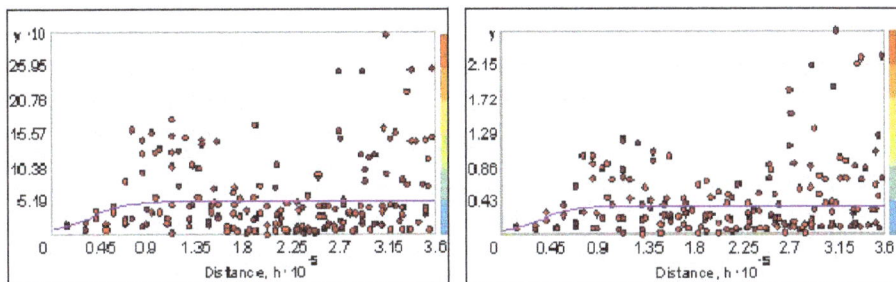

Figure 2: Best fitted semi-variogram model assuming constant trend

Table 1 illustrates the best fitted semi-variogram model parameters. It also illustrates a measure for the degree of spatial dependence given by the share of the variability due to spatial dependence (sill) to the total variability (sill + nugget). Examining Table 1, it is found that both the WWQI and the DWQI exhibit strong spatial dependence between the sampling points along the Nile up to a distance of 87 and 81 km., respectively. The cross validation statistics (Table 2) indicate that the selected semi-variogram models and the associated parameters are reasonable. They have the smallest RMSE, the MStE are closer to zero and the RMSStE are closer to one.[3]

Variable	Model	Number of lags	Nugget C_o	Sill C_1	Range a	$\frac{Sill}{Sill+Nugget}$
WWQI	Gaussian	18	0.09	0.42	87 km.	0.82
DWQI	Gaussian	18	0.06	0.29	81 km.	0.83

Table 1: Best semi-variogram model assuming constant trend

Variable	ME	RMSE	MStE	RMSStE
WWQI	-2.265	50.86	-0.088	0.9452
DWQI	-3.792	114.8	-0.058	1.001

Table 2: Cross validation statistics of ordinary kriging assuming constant trend

The generated WWQI map depicted in Figure 3 shows excellent water quality at Upper Egypt, except at Asuit where good and poor water qualities are noticed. Also, it shows that the water quality varies between excellent and good at Middle Egypt. Noticeably, the water is of poor quality at Beni-Suef. Generally, Lower Egypt suffers from poor water quality, especially at El-Sarw Drainage. Indeed, the water quality is very poor in Rosetta Branch, especially in Kafr El-Zayat. The contour map of prediction errors (Figure 3) indicates that the prediction errors of the points located in Middle and Lower Egypt are considerably large relatively to those sited in Upper Egypt. Moreover, it shows that the prediction standard errors are relatively smaller around the sampling points than in areas without sampling points.

The DWQI map (Figure 4) represents the spatial distribution of the drinking water quality. The water of the Nile is of good quality for drinking in Upper Egypt, except in Asuit. Poor drinking water qualities are depicted from the map in Greater Cairo and Lower Egypt. An interesting point is that the DWQI is above 100 at all the drinking water intakes along the Nile in Middle and Lower Egypt, indicating poor drinking water qualities. The contour map of prediction errors (Figure 4) shows that the prediction errors are large, in general. However, the prediction errors of the points located in Middle and Lower Egypt are larger compared with those sited in Upper Egypt. This result is due to the presence of extreme outliers. Moreover, the prediction errors, as expected, are larger in Luxor, Qena and Rosetta branch due to the shortage of sampling points.

Figure 5 illustrates the percentage contribution of each polluting variable to the drinking water quality index, which could help in determining the cause of water pollution. The figure

---

[3]This model is the best fitted model as compared to all other semi-variogram models.

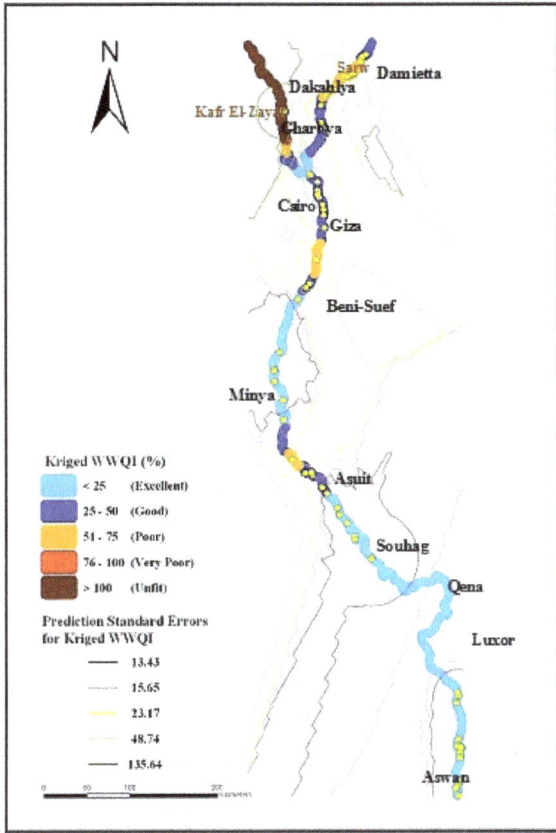

Figure 3: Surface generation of the WWQI and contour map of the prediction standard errors for kriged WWQI

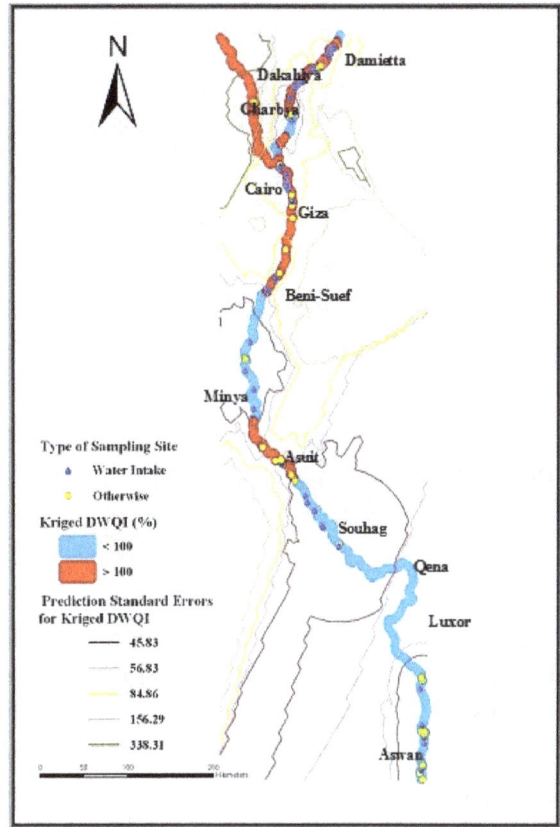

Figure 4: Surface generation of the DWQI and contour map of the prediction standard errors for kriged DWQI

shows that the unfit water for drinking use dominant in Asuit are attributed to the high levels of Ammonia (NH_4). As well as, the poor drinking water quality in Beni-Suef and Greater Cairo are due to the high levels of heavy metals ($Mn + Fe$). Additionally, the figure indicates that the high prevalence of biological oxygen demand (BOD) has a great effect on the water quality. This means that the organic waste is problematic in Aswan, Sohag, Menya and Gharbeya. A major type of organic waste is human waste, which usually involves significant human pathogens creating a health hazard (Rahman and Hossain 2008).

5. Discussion

The Nile pollution becomes a pressing national issue in Egypt. This is due to the accelerated population growth and economic activities, which impose a heavy burden on the viability of the Nile water quality. Accordingly, this study aims to assess the Nile water quality in Egypt at "February 2008" using spatial statistical analysis. In fact, spatial interpolation techniques facilitate the identification of highly polluted areas. This in turn will aid decision makers in adopting appropriate policies to combat and prevent the Nile water pollution.

After exploring data properties and resorting to ordinary kriging to spatially interpolate the

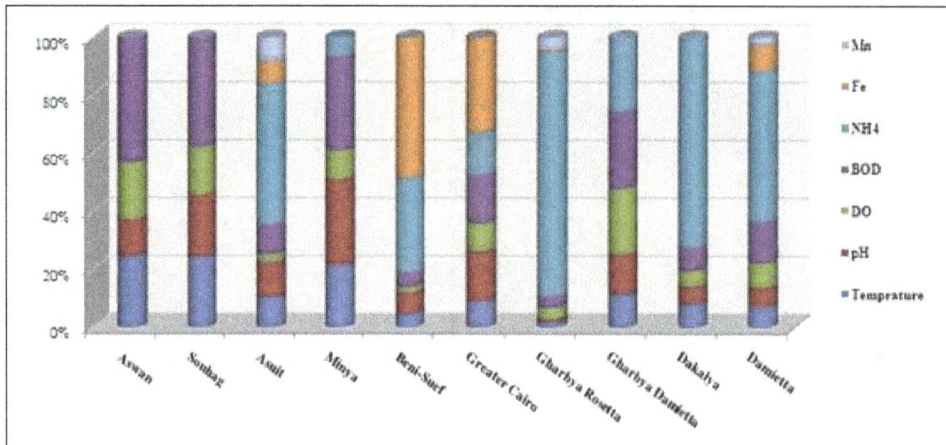

Figure 5: Surface generation of the DWQI and contour map of the prediction standard errors for kriged WWQI

water quality along the Nile in Egypt, cross-validation technique proved that the Gaussian models are the best fitted semi-variogram models describing the spatial variability of both indices, WWQI and DWQI. Spatial interpolation is performed on the bases of these models and surfaces are generated to map the spatial distribution of the water quality.

It can be concluded from the results of the study, in conjunction with the findings of other relevant studies and information, that:

- Upper Egypt has excellent water quality, except Asuit which is characterized by water quality varying between good and poor. Poor water quality in Asuit is attributed to agriculture discharges and fertilizers factories situated on its river banks (CEMSWE 2008);

- Water quality varies between poor and good in Middle Egypt and along Damietta branch. Very poor water quality is observed at the Rosetta branch due to accumulated industrial discharges into the river. These discharges originate from the industrial complex of pesticides and fertilizers located at Kafr El-Zayat and some factories located in Greater Cairo. Besides agricultural discharges stemming from the agricultural drainages in Beni-Suef (CEMSWE 2008);

- Excluding Asuit, good water suitable for drinking after applying simple physical treatments and disinfection in the water intake plants is situated at Upper Egypt;

- Unfit water for drinking, dominant at Asuit and along Rosetta and Damietta branches, is attributed to the high levels of Ammonia (NH_4) originating from the agricultural drainages and the wastes of the pesticides and fertilizers factories. The poor drinking water quality at Beni-Suef and Greater Cairo is due to high levels of heavy metals (Mn + Fe)(CEMSWE 2008). Therefore, simple physical and chemical treatment with disinfection is not adequate and intensive physical and chemical treatments with disinfection are the most suitable options for the water intake plants in these areas. Moreover, strict water quality standards should be imposed on drains and wastewater discharges; and

- The BOD has a significant negative effect on water quality in Aswan, Sohag, Menya and Gharbeya. This can be attributed to the bad municipal sewage network covering those governorates (CEMSWE 2008). These high BOD levels reduce the ability of water to sustain aquatic life, inducing negative impact on the ecosystem and fisheries.

In light of the results and the experience and knowledge acquired during field practice, the following recommendations may be useful for future research:

- The sampling points should be increased to enhance the accuracy of estimation in areas with few or without sampling points;

- Taking the direction of the water flow into account would produce better results since, in a river system, sampling site Y is only affected by upstream sampling site X; and

- Carrying out further studies which highlight seasonal distribution and spatio-temporal analysis and emphasize the concept of the dynamic transport of pollutants loadings.

Acknowledgements

The authors would like to acknowledge the partial financial support of the Information and Decision Support Center (IDSC). They also thank professor Hanaa Kheir-El-Din and Ragaa El-Wakil for revising and editing earlier version of the paper. In addition, they thank professor Reda Mazloum, professor Mohammed Ismail, the participants of the 22nd annual conference of The International Environmetrics Society (TIES 2012) and the anonymous referee for his valuable comments that have substantially improved the paper. The usual disclaimer applies.

References

Abou-Ali H, El-Ayouti A (2012). "Nile Water Heterogeneity in Egypt: Assessment Using Spatial Analysis." *Information and Decision Support Center Working Paper Series*, **no.29**.

Bordalo A, Nilsumranchit W, Chalermwat K (2001). "Water Quality and Used of the Bnagpakong River." *Water Research*, **35**, 3635–3642.

CEMSWE (2008). "The Annual Report of the Results of the National Network for monitoring the water Pollutants of the Nile River and its Branches in 2008." *Technical report*, Central Department of Environmental Affairs - Center of Environmental Monitoring and Studies of the Working Environment (CEMSWE) - Ministry of Health, Egypt.

Chang H (2008). "Spatial Analysis of Water Quality Trends in the Han River Basin, South Korea." *Water Research*, **42**, 3285–3304.

Donia N, Farag H (2010). "A Waste Load Model Analysis for El-Noubariya Canal Drinking Water Abstraction." In *Proceedings of the Fourthenth International Water technology Conference*.

Dowd P (2003). "Testing for Constant Spatial Mean using the Global D-Statistic." *Computers & Geosciences*, **29**, 1057–1068.

Elmahdi A, Afify A, Abdin A (2008). "Development of a GIS-based Decision Support Tool and Assessment of Nile River Water Quality." *International Journal of Water*, **4**, 55–68.

Flipo N, Jeannee N, Poulin M, Even S, Ledoux E (2007). "Assessment of Nitrate Pollution in the Grand Morin Aquifers (France): Combined Use of Geostatistics and Physically based Modeling." *Envionmental Pollution*, **146**, 241–256.

Fortin MJ, Dale MT (2005). *Spatial Analysis - A Guide For Ecologists*. First edition. Cambridge University Press, U. K.

French J (2005). *Exploring Spatial Correlation in Rivers*. Master's thesis, Department of Statistics, Colorado State University, USA.

Haining R (1990). *Spatial Data Analysis in the Social and Environmental Sciences*. First edition. Cambridge University Press, U. K.

Hamad S (2008). *Spatial Analysis of Groundwater Level and Hydrochemsitry in the South Al Jabal Al Akhdar area Using GIS*. Master's thesis, Centre of Geoinformatics, Slazburg University, Austria.

Johnston K, Hoef J, Krivoruchko K, Lucas N (2001). *Using ArcGIS Geostatistical Analyst. GIS by ESRI*. Illustrated edition. ESRI, New York.

Ouyang Y, Higman J, Thompson J, O'Toole T, Campbell D (2001). "Charecterization and Spatial Distribution of Heavy Metals in Sediment from Cedar and Ortega Rivers Subbbasin." *Contaminant Hydrology*, **54**, 19–35.

Rahman S, Hossain F (2008). "Spatial Assessment of Water Quality in Peripheral Rivers of Dhaka City for Optimal Relocation of Water Intake Pointl." *Water Resources Management*, **22**, 377–391.

Sarangi A, Madramootoo CA, Enright P (2006). "Comparison of Spatial Variability Techniques for Runoff Estimation from Canadian Watershed." *Biosystems Engineering*, **95**, 295–308.

Tiwari TN, Mishra M (1985). "A Preliminary Assignment of Water Quality Index of Major Indian Rivers." *Indian Journal of Environmental Protection*, **5**, 276–279.

Affiliation:

Corresponding Author: **Amira El-Ayouti**
Department of Statistics, Faculty of Economics and Political Science, Cairo University
Cairo, Egypt.
E-mail: `amiraelayouti@msn.com`, `a.ayouti@feps.edu.eg`
Hala Abou-Ali
Department of Economics, Faculty of Economics and Political Science, Cairo University
Cairo, Egypt.

7

Spatio-temporal evolution modeling of environmental and natural phenomena

Jorge Mateu
University Jaume I, Castellón

Abstract

This short paper introduces the special issue containing six selected papers coming from the International Workshop on Spatio-temporal Modeling (METMA V) held in Santiago de Compostela (Spain), from 30 June to 2 July 2010.

Keywords: Anisotropy; Average temperatures; Spatial clustering; Functional processes; Integrated nested Laplace approximation; Spatial variability; Spatio-temporal modeling; Two-stage Bayesian approach; Wavelet analysis.

1. Introduction

In recent years, spatio-temporal modeling has become one of the most interesting and, at the same time, challenging research areas of natural and environmental sciences. The relevant literature is growing fast and along directions that range from theoretical and methodological developments to real world applications. Spatio-temporal systems modeling involves the synthesis of a rich interdisciplinary body of knowledge for which it is necessary to establish a solid theoretical foundation and a science-based methodology with both researchers and practitioners in mind.

In this context, the biannual International Workshop on Spatio-temporal Modeling (METMA) has reached its fifth edition, and has been held in Santiago de Compostela (Spain), from 30 June to 2 July 2010. With the purpose of promoting the development and application of spatio-temporal statistical methods in different fields related to environmental sciences, METMA V contributions gathered the latest advances in statistical methodology illustrated with environmental data. With more than 150 participants from 10 different countries, senior and junior researches, who attended the invited and contributed sessions (with a no-parallel session system, which favored the exchange of ideas and experiences), this international workshop has become a cornerstone in the worldwide meetings related to this topic, while keeping the maximum number of participants within a very reasonable size.

We would like to express our gratitude and thanks to all the members of the local organizing and scientific committee. We also would like to thank all the participants for their contributions to the spatio-temporal field in its wide sense: geostatistics, point processes, and lattice data. Two Special Issues have arisen from this workshop, one published in *Environmetrics*, and the other one here in *Journal of Environmental Statistics*. The complete list of papers presented at the workshop and any particular information are posted at: http://eio.usc.es/pub/metma/index.php?lang=en.

In light of the above considerations, the articles of this Special Issue have been carefully selected to present a variety of conceptual frameworks, powerful methods and comprehensive techniques that address a number of interesting problems in environmental, health, social and medical sciences. Bayesian hierarchical spatio-temporal models are considered in Saez *et al.* and Jin *et al.* Integrated nested Laplace approximations within the Bayesian approach for replicated data in time are used in Illian *et al.* Wavelet analysis for detecting and comparing trends, investigating spatial heterogeneity and periods of significant variability in non-stationary environmental time series is considered in Franco-Villoria *et al.* Anisotropy and spatial clustering effects are analyzed by Kelly. Finally, a functional approach for spatio-temporal strong dependence processes is presented by Frias and Ruiz-Medina.

Assessing the spatial and temporal variability of temperatures, through describing the mean function as well as the spatio-temporal covariance structure of maximum and minimum daily temperature data is the core aim in Saez *et al.* in their paper entitled *Space-time interpolation of daily air temperatures*. The authors propose an estimation of the long-term trend of maximum and minimum temperatures, model the spatio-temporal variability of temperatures, and interpolate the spatial temperatures at any given time. Long-term trend, annual harmonics and winds are considered as explanatory variables of the mean function. The parameters associated with these variables are allowed to vary between stations and within each year. The temporal autocorrelation is controlled through ARMA models, and spatio-temporal models are built as Bayesian hierarchical models with two stages following the integrated nested place Laplace approximation (INLA) for Bayesian inference. The spatial parameters are also allowed to vary with time.

A statistical method for modeling ground-level ozone concentration in a given region is developed and presented in Jin *et al.*, where an environmental network design problem is also explored. By applying hierarchical Bayesian spatio-temporal modeling, a conditional predictive distribution over a set of grid points is obtained. In terms of an entropy criterion, the environmental network design problem is solved using the obtained predictive distributions. Model evaluation is also provided.

As it is known and has become popular in the very recent years, integrated nested Laplace approximation (INLA) provides a fast and exact approach to fitting complex latent Gaussian models which comprise many statistical models in a Bayesian context. Illian *et al.* discuss how a joint log Gaussian Cox process model may be fitted to independent replicated point patterns. They illustrate the approach by fitting a model to data on the locations of muskoxen herds in Zackenberg valley, Northeast Greenland and by detailing how this model is specified within the R-interface R-INLA.

A wavelet analysis is presented in Franco-Villoria *et al.* as a possible method for detecting and comparing trends, investigating spatial heterogeneity and periods of significant variability in non-stationary environmental time series. Their results confirm a difference in river flow maxima between regions due to changes in the seasonal patterns that may be linked to external climatic drivers, including the North Atlantic Oscillation and the Atlantic Multidecadal Oscillation. Such influences (which act on several time scales, the principal one being annual) are not constant and vary both temporally and spatially, with a possible catchment size effect, highlighting the importance of assessing flood risk at a regional level.

A practical problem on examining the spatial association of bovine TB in cattle herds using data from Ireland is presented in Kelly. Badgers, a protected species, have been implicated in the spread of the disease in cattle. Current disease control policies include reactive culling (in response to TB outbreaks) of badgers in the index and neighbouring farms. The author accounts for possible anisotropy. Changes in the spatial association over two time periods are also examined. The results have direct implications for establishing scale and direction in reactive culling. They are also important regarding the evaluation of vaccines for badgers and cattle.

Filtering and parameter estimation are addressed in the context of spatio-temporal strong dependence processes. A functional parametric observation model is fitted to the spectral sample information. Large dimensional spectral data sets, displaying high local singularity, are then processed in this functional setting. Thresholding techniques are first applied for removing noise generated from measurement spectrometer device. Spatio-temporal long-range dependence model fitting is then achieved by applying linear regression in the log-wavelet domain.

We hope the reader finds this selection of papers in the spatial and spatio-temporal context useful, comprehensive and inspiring. In that case, the METMA V atmosphere will have been honestly translated into this volume. We finally would like to thank the authors for their outstanding research contributions and, in particular, Prof. Rick Schoenberg (Editor-in-Chief of Journal of Environmental Statistics) for his encouragement and support in preparing this Special Issue.

Affiliation:

J. Mateu.

Statistics and Operations Research, Department of Mathematics, University Jaume I, E-12071 Castellón, Spain.

8

Statistical Climate-Change Scenarios

Jan R. Magnus
Vrije Universiteit Amsterdam

Bertrand Melenberg
Tilburg University

Chris Muris
Simon Fraser University

Martin Wild
ETH, Zürich

Abstract

We report on climate projections generated by a simple model of climate change. The model captures the effects of variations in surface solar radiation, using information over the period 1959–2002 available from observational records from the Global Energy Balance Archive (GEBA), as well as increases in greenhouse gases on surface temperature. The model performs well with respect to observational data, and is simple enough to admit a rigorous statistical analysis. This allows us to quantify the uncertainty associated with estimated parameter values using observational data only. Our method immediately leads to estimates with associated confidence intervals, which can be translated into confidence intervals for climate projections. In particular, we construct probabilistic climate projections using standard scenarios for carbon dioxide and sulphur dioxide emissions.

Keywords: global warming, dimming, aerosols, projections.

1. Introduction

Changes in the concentrations of greenhouse gases and aerosols are the two most important drivers of man-made climate change. Modeling the processes through which these two variables affect our climate is therefore an essential ingredient of any climate model. Using scenario analysis, these climate models are then used to provide policy makers with climate projections, conditional on hypothesized changes in greenhouse gas and aerosol emissions.

Uncertainty plays an important role in modeling and projecting climatic change; see, for example, Andreae et al. (2005), Stainforth et al. (2005), and Roe and Baker (2007). Given a climate model, uncertainty about the parameter values leads to uncertainty in the implied climate projections. While consensus exists on the values of some parameters, there is much uncertainty on many key parameter values, for example the value of climate sensitivity and the aerosol effect; see, for instance, Schwartz et al. (2010) and Knutti and Plattner (2012).

Our aim is to formulate a climate model, which is simple enough to allow a rigorous statistical analysis, but not so simple that it ignores key climate ingredients. The simplicity is essential for our purpose, because it implies that we can estimate the parameters of the climate model with conventional statistical methods. This has the advantage that we can quantify the uncertainty associated with the estimated parameter values. In other words, we let observational data tell us how confident we can be about the parameter values for our model, and how well the model fits the data. We then translate this uncertainty into confidence intervals for projections of future climate under various scenarios.

The starting point is a model recently proposed in Magnus et al. (2011) who attempted to disentangle the counteracting effects on surface temperature of the observed reductions of surface solar radiation and of increases of greenhouse gases. The parameters of this model are estimated using observational data over the period 1959–2002 obtained from the Global Energy Balance Archive (GEBA), and surface temperature and CO2 concentration data. In order to conduct scenario analysis, we augment this model in two directions. First, we provide carbon and aerosol models, linking emissions to concentrations. Second, we develop a model that allows us to distinguish between model error and measurement error. We then estimate the parameters in these models, quantify the uncertainty associated with the estimates, and apply our model to typical climate scenarios.

The simplicity of the model allows us to quantify the uncertainty associated with estimated parameter values using observational data only. As a consequence, our method immediately leads to estimates with associated confidence intervals, which are translated into confidence intervals for climate projections. In this way, our statistical approach differs from the probabilistic approach employed by the IPCC; see Meehl et al. (2007). The latter approach makes use of models, based on physical principles, that are more advanced and closer to reality. Such advanced models are typically characterized by many parameters, and as such, too complex to be estimated directly, solely on the basis of observational data. Instead, they have to be validated using indirect methods using implied relationships, for instance, by confronting model-based values of the climate sensitivity with empirical estimates of this quantity; see, for instance, Knutti and Hegerl (2008) and references therein. Probabilistic climate projections using advanced and complex models can then be obtained by varying the parameters over such indirectly validated values. Our observationally-based approach can be seen as a *direct* way to validate this procedure: our relationships, estimated (in-sample) using observational data, are directly 'extrapolated' into (out-of-sample) climate projections, given typical climate scenarios. Finding substantial differences between the two approaches would require at least that this difference is not caused by the indirect validation. However, our findings do not invalidate the IPCC outcomes.

The remainder of this paper is organized as follows. The energy balance is discussed in Section 2. In Section 3, we introduce uncertainty in the energy balance equation, leading to a statistical model, and describe the data used to estimate this statistical model. In Section 4 we decompose this uncertainty into process risk and measurement error. In Section 5 we provide a simple link between emissions and concentrations. This completes the model. Section 6 briefly describes the three scenarios that we consider. Section 7 contains the results of our scenario and validation analysis, and Section 8 concludes. A data appendix contains additional information about the emission data used for estimating the link between emissions and concentrations proposed in Section 5.

2. The energy balance

Our starting point is the energy balance equation,

$$c \cdot \frac{d\text{TEMP}_t}{dt} = \text{EB}_t, \tag{1}$$

where TEMP_t denotes the surface temperature at time t (measured in degrees Celsius), EB_t is the energy balance, c the heat capacity, and $d\text{TEMP}_t/dt$ the derivative of TEMP_t with respect to time t. Decomposing the energy balance as in Wild et al. (2004), gives

$$\text{EB}_t = \text{SW}_t^{\text{abs}} + \text{LW}_t^{\text{down}} + \text{LW}_t^{\text{up}} + \text{SH}_t + \text{LH}_t + \text{GH}_t + \text{M}_t, \tag{2}$$

where SW^{abs} is the absorbed shortwave radiative flux, LW^{down} and LW^{up} are the downward and upward longwave radiative fluxes, SH and LH are the sensible and latent heat fluxes, GH is the ground heat flux, and M is the energy flux used for melt.

Wild et al. (2004) analyzed the change in energy fluxes over the period 1960–1990. For their purpose it sufficed to consider the stationary perturbation surface temperature, so that $\Delta \text{EB}_t = 0$, where Δ denotes a change per unit of time: $\Delta x_{t+1} = x_{t+1} - x_t$. In contrast, we implement (1) using annual data from observational sites (which include measures of surface solar radiation along with conventional synoptic weather information). We describe these observational data in detail in the next section. To transform the energy balance equation to annual changes in temperature we integrate (1) over a one-year period,

$$c \int_t^{t+1} \frac{d\text{TEMP}_\tau}{d\tau} \, d\tau = \int_t^{t+1} \text{EB}_\tau \, d\tau, \tag{3}$$

leading to the approximation

$$c \left(\text{TEMP}_{t+1} - \text{TEMP}_t \right) \approx \text{EB}_t. \tag{4}$$

The approximation will be more accurate when we measure the energy balance EB_t as a one-year average, because seasonal effects are then balanced out, and this is what we shall do in our empirical analysis. Assuming equality in (4), we write the equation in differences (over one-year periods),

$$c\Delta \text{TEMP}_{t+1} = \Delta \text{EB}_t + c\Delta \text{TEMP}_t. \tag{5}$$

We next specify the energy balance term ΔEB_t. Given (2) and assuming that changes in the ground heat flux GH_t and the energy flux used for melt M_t are negligible (Wild et al., 2004, Table 1; Wild et al., 2008, Table 1), we obtain

$$\Delta \text{EB}_t = \Delta \text{NSR}_t + \Delta \text{SH}_t + \Delta \text{LH}_t,$$

where NSR denotes the net surface radiation

$$\text{NSR} = \text{SW}^{\text{abs}} + \text{LW}^{\text{down}} + \text{LW}^{\text{up}}.$$

Following Magnus et al. (2011) we parameterize the terms in the net surface radiation equation as

$$SW_t^{abs} = (1 - \alpha_1)RAD_t,$$
$$LW_t^{down} = \alpha_2 + \alpha_3 \log(CO2_t),$$
$$LW_t^{up} = \alpha_4 + \alpha_5 TEMP_t,$$

and this leads to

$$\Delta NSR_t = (1 - \alpha_1)\Delta RAD_t + \alpha_3 \Delta \log(CO2_t) + \alpha_5 \Delta TEMP_t. \tag{6}$$

In these equations, RAD_t stands for the solar surface radiation, that is the solar radiation reaching the Earth's surface (measured in Watts per meter squared), and $CO2_t$ stands for the carbon dioxide concentration (measured in parts per million by volume). To close the model we need to parameterize the latent and sensible heat fluxes. We shall assume that changes in these fluxes are proportional to changes in the net surface radiation, that is,

$$\Delta SH_t = \alpha_6 \Delta NSR_t, \qquad \Delta LH_t = \alpha_7 \Delta NSR_t. \tag{7}$$

Then, substituting (6) and (7) into (5), we find

$$\Delta TEMP_{t+1} = \beta_1 \Delta TEMP_t + \beta_2 \Delta RAD_t + \beta_3 \Delta \log(CO2_t), \tag{8}$$

where the β's are linear combinations of the α's.

3. The statistical model

For each site in our observational data set, we apply this model for every year for which data is available. Because of neglected terms, approximation errors in the parametrization, and possible measurement errors, we allow for stochastic error terms, $\Delta u_{i,t+1}$, for each observational site for each time period, and a time-specific temperature change $\Delta \lambda_{t+1}$ that is common to all weather stations:

$$\Delta TEMP_{i,t+1} = \beta_1 \Delta TEMP_{i,t} + \beta_2 \Delta RAD_{i,t} + \Delta \lambda_{t+1} + \Delta u_{i,t+1}. \tag{9}$$

The time-specific temperature change depends on global average temperature, global average surface solar radiation, carbon dioxide concentration, and an additive stochastic error term:

$$\Delta \lambda_{t+1} = \gamma_1 \Delta \overline{TEMP}_t + \gamma_2 \Delta \overline{RAD}_t + \gamma_3 \Delta \log(CO2_t) + \eta_{t+1}, \tag{10}$$

where global averages are denoted by a horizontal line over the variable. This specification is the same as Equations (10) and (11) in Magnus et al. (2011), though obtained via a different route.

To estimate Equations (9) and (10), we collected monthly observations on three variables:

- temperature (TEMP), the average temperature in degrees Celsius (°C) at the surface (the near-surface temperature), as expressed as anomalies from a base period (1960–1990). *Source:* Climatic Research Unit (CRU TS 2.1) at the University of East Anglia in the UK (Mitchell and Jones, 2005), see http://www.cru.uea.ac.uk;

- surface solar radiation (RAD), the amount of sunlight ('global solar irradiance') reaching the Earth's surface, measured in Watts per meter squared (Wm^{-2}). *Source:* Global Energy Balance Archive (GEBA) (Gilgen and Ohmura, 1999);

- carbon dioxide concentration (CO2), measured in parts per million by volume (ppmv). *Source:* Mauna Loa Observatory (MLO) in Hawaii, see http://www.mlo.noaa.gov/

Data from the three sources are linked through their locations. The radiation data are incomplete. We circumvent a potential sample selection problem by considering temperature changes rather than temperature levels. We use annual data rather than monthly data to avoid problems of seasonal adjustment. This provides us with $N = 1337$ observational sites over a period of $T = 44$ years (1959–2002).

The parameters are estimated using the generalized method of moments (GMM) approach using appropriate moment restrictions, following Arellano and Bond (1991), Blundell and Bond (1998), and Magnus et al. (2011). The resulting parameter estimates and their standard errors are

$$\Delta\text{TEMP}_{i,t+1} = \underset{(0.0046)}{0.9063}\,\Delta\text{TEMP}_{i,t} + \underset{(0.0008)}{0.0087}\,\Delta\text{RAD}_{i,t} + \Delta\lambda_{t+1} + \Delta u_{i,t+1} \tag{11}$$

and

$$\Delta\lambda_{t+1} = \underset{(0.1839)}{-0.8235}\,\Delta\overline{\text{TEMP}}_t + \underset{(0.0219)}{0.0614}\,\Delta\overline{\text{RAD}}_t + \underset{(2.3958)}{10.6955}\,\Delta\log(\text{CO2}_t) + \eta_{t+1}. \tag{12}$$

Averaging over (11) and combining with (12) then leads to

$$\Delta\overline{\text{TEMP}}_{t+1} = \underset{(0.1839)}{0.0828}\,\Delta\overline{\text{TEMP}}_t + \underset{(0.0219)}{0.0701}\,\Delta\overline{\text{RAD}}_t + \underset{(2.3958)}{10.6955}\,\Delta\log(\text{CO2}_t) + \eta_{t+1} \tag{13}$$

where we have assumed that $\Delta\bar{u}_{t+1} = 0$. The estimates in (11) are much more accurate than those in (12). There is not much loss, therefore, if we calculate the standard errors in (13) based on the standard errors in (12) alone.

In the sequel we shall write (13) briefly as

$$\Delta\overline{\text{TEMP}}_{t+1} = \mu_t + \eta_{t+1}, \tag{14}$$

with μ_t representing the 'systematic' part, depending on lagged temperature, surface solar radiation, and carbon dioxide concentration, and η_{t+1} the so-called 'idiosyncratic' part of the average temperature change, that is the change in temperature not captured by the systematic part.

4. Specification of η_{t+1}

We wish to apply (14) to forecasting and scenario analysis. Since the uncertainty in our model is driven by observational data, such a scenario analysis will give a realistic view of the degree of uncertainty about future climate change. There are, however, two problems that need to be resolved before we can attempt this analysis. First, we need to specify how the errors η_{t+1} are generated. Second, inputs to the model are carbon dioxide concentration CO2 and surface solar radiation RAD, while the scenarios are in terms of emissions of CO2 and SO2, so we need a link between these emissions and the inputs to the model. We deal with each issue in turn. The current section discusses the specification of the errors η_{t+1}, and the next section proposes simple models which transform carbon dioxide emissions to concentrations and sulphur dioxide emissions to surface solar radiation.

The idiosyncratic error term η_{t+1} in (14) consists of idiosyncratic process risk (representing natural variations in the average temperature change not captured by the systematic part μ_t) and measurement error. We shall identify this measurement error, so that it does not enter into our climate projections. To do so, we compare a time series based on our data set (the CRU TS 2.1 series) to the global means of the CRUTEM3v data set (see Brohan et al., 2006). These data sets overlap in terms of the basic temperature series, but are different in their coverage and how they are constructed. The coverage of our data set is restricted to the 1337 observational sites in our data set, while the CRUTEM3v uses data on all observational sites available for the period under consideration. To arrive at a global mean temperature, we use a simple averaging method, while a more sophisticated algorithm was used for CRUTEM3v. In terms of temperature differences there is no selection bias (see Magnus et al., 2011), but there might be a difference in the measurement error of these two temperature series. We therefore propose the following set-up, where the superscript C refers to the CRUTEM3v temperature series:

$$\Delta\overline{\text{TEMP}}_{t+1} = \mu_t + \eta_{t+1} = \mu_t + \eta_{t+1}^* + \epsilon_{t+1},$$
$$\Delta\overline{\text{TEMP}}_{t+1}^C = \mu_t + \eta_{t+1}^C = \mu_t + \eta_{t+1}^* + \epsilon_{t+1}^C.$$

Here, η_{t+1}^* denotes the idiosyncratic process risk, and ϵ_{t+1} and ϵ_{t+1}^C represent the measurement error in the temperature series used in the estimation procedure and the CRUTEM3v temperature series, respectively. The systematic part μ_t and the idiosyncratic process risk η_{t+1}^* are the same for both temperature series, while each temperature series is assumed to have its own measurement error.

Our aim is to retrieve the idiosyncratic process risk η_{t+1}^*. We cannot, however, retrieve three error terms (η_{t+1}^*, ϵ_{t+1}, ϵ_{t+1}^C) from two observed temperature series. Therefore we propose, in addition, that the two measurement errors are proportional, that is,

$$\epsilon_{t+1}^C = f \times \epsilon_{t+1},$$

for some fraction f, constant over time. Given f, we can identify ('solve') η_{t+1}^*.

We choose f such that the sample correlation between the idiosyncratic process risk and the measurement error vanishes in both series. The condition $\text{COV}(\eta_{t+1}^*, \epsilon_{t+1}) = 0$ leads to the estimate $\widehat{f} = 0.0826$, so that

$$\eta_{t+1}^* = \frac{\eta_{t+1}^C - \widehat{f}\eta_{t+1}}{1 - \widehat{f}} = 1.09\,\eta_{t+1}^C - 0.09\,\eta_{t+1}.$$

The low value of $\widehat{f}$ shows that the measurement error in the CRUTEM3v temperature data series is smaller than in the data that we use (the CRU TS 2.1 series). This is as expected, as our data consist of just one of the series that are used for computing the CRUTEM3v series.

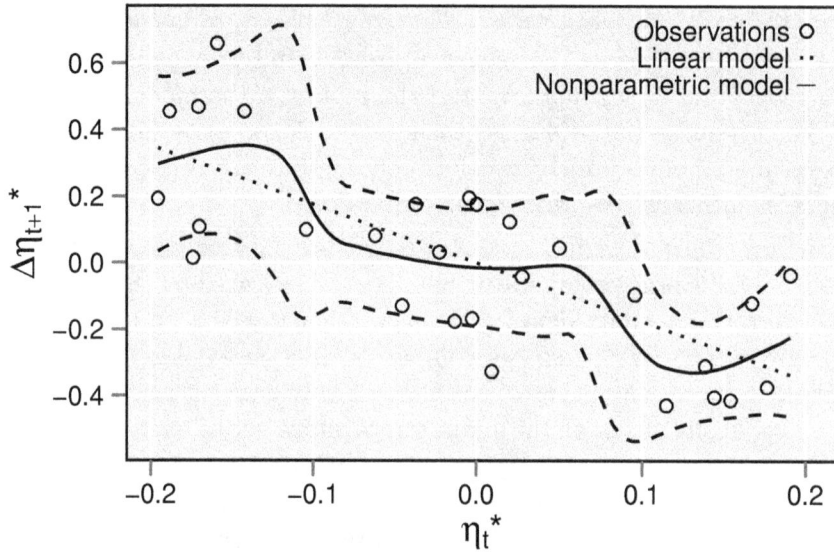

Figure 1: Justifiability of the Ornstein-Uhlenbeck process

Given this estimated value of f, the total estimated variance of η_{t+1} (0.1589) is equal to the sum of the estimated variance of η_t^* (0.0505) and the estimated variance of ϵ_{t+1} (0.1084).

Now that we have data on the idiosyncratic process risk η_{t+1}^*, we specify the underlying process as follows. Following Majda at al. (2001), Vallis et al. (2004), and Padilla et al. (2011), we model η_{t+1}^* as an Ornstein-Uhlenbeck type process:

$$\Delta\eta_{t+1}^* = -\alpha\eta_t^* + \zeta_{t+1}, \qquad \zeta_{t+1} \overset{iid}{\sim} \left(0, \sigma_\zeta^2\right). \tag{15}$$

The dotted line in Figure 1 shows the regression for $\widehat{\alpha} = -1.7582$, restricted to the range $\eta_t^* \in [-0.2, +0.2]$. The figure also plots the nonparametric estimate of g in

$$\Delta\eta_{t+1}^* = g(\eta_t^*) + \omega_{t+1}, \qquad E\left(\omega_{t+1}|\eta_t^*\right) = 0,$$

together with the corresponding 95% uniform confidence band, where we apply Härdle and Linton (1994, eq. (29)), using the quartic kernel and bandwidth chosen according to Silverman's rule of thumb with some undersmoothing. (Without this undersmoothing, the nonparametric graph would be close to linear.) The nonparametric regression provides empirical evidence in support of modeling the temperature error by means of an Ornstein-Uhlenbeck type process. The resulting estimated variance of ζ_{t+1} is $\widehat{\sigma}_\zeta^2 = 0.0347$. The subsequent results are based on this model for the error process.

5. Emissions and concentrations

Climate scenarios are usually specified in terms of emissions, not in terms of concentrations. For example, a scenario may prescribe a specific increase in carbon dioxide emissions or a decrease in sulphur dioxide emissions. Our energy balance model (9)–(10) is formulated in terms of CO2 concentration and surface solar radiation. Hence, we need auxiliary models that transform carbon dioxide emissions into carbon dioxide concentrations, and sulphur dioxide emissions into levels of surface solar radiation. Although there are also, for example, SRES CO2 concentration scenarios, we prefer to use our own transformations from emissions to concentrations, in order to be able to quantify the uncertainties arising from the use of observational data.

The auxiliary models presented in this section follow the general philosophy of the paper in that they are simple models analyzed in a statistically rigorous way. They are sufficiently simple that they can be estimated using conventional statistical techniques, so that we can incorporate the uncertainty in the estimates of the model parameters in our climate projections. This allows us to use the augmented model to convert scenario data — CO2 and SO2 emissions — into probabilistic climate projections that take the uncertainty into account at every step of the modeling process. The emission data used for estimation are described in the data appendix.

We estimate a linear regression between the change in carbon dioxide concentrations ($\Delta CO2$) and CO2 emissions (CE):

$$\Delta CO2_t = 0.0993 + 0.2436\,CE_t + v_t. \tag{16}$$
$$(0.0404)\quad(0.0521)$$

In addition, we estimate a linear regression linking global average surface solar radiation ($\overline{RAD}$) to aggregate SO2 emission (SE).

$$\overline{RAD}_t = 59.1272 - 15.4063\,\log(SE_t) - 0.2872\,(t - 1958) + w_t \tag{17}$$
$$(7.4184)\quad(1.7954)\qquad\quad(0.0169).$$

Equation (17) does not aim to provide a structural model for average surface solar radiation in terms of aggregate SO2 emission. The equation simply represents an *empirical* link, based on past data, for the purpose of transforming SO2 emission data into average surface solar radiation data. We shall interpret the time trend as reflecting growth, which, measured by world GDP in logarithmic terms, is very close to linear as a function of time. Growth likely results in emissions not captured by SO2 emissions. These omitted emissions might affect the surface solar radiation directly, like SO2 emissions, or indirectly, such as greenhouse gases, including CO2, that influence the surface solar radiation via increasing cloud thickness, as suggested by Tselioudis and Rossow (1994). For a review of the literature on the relationship between aerosols (such as sulphur dioxide) and surface solar radiation, see Wild (2009). The data used to estimate these models are described in the data appendix.

The energy balance model (9)–(10) resulting in (13), together with the error specification (15) and the two auxiliary equations (16) and (17) constitute our augmented climate model. Its performance is illustrated in Figures 2 and 3.

The left panel of Figure 2 describes the in-sample fit of the carbon model (16), while the right panel of Figure 2 presents the aerosol-surface solar radiation model (17), with the level of average surface solar radiation measured in Wm^{-2} in deviation of the average level of surface solar radiation in 1959. The small circles are the actual observations, the solid curves the in-sample forecasts (starting at the beginning of the sample), and the dashed curves are uncertainty bands. We include both parameter and process uncertainty, using 5000 simulations. In each simulation, parameters are drawn from joint asymptotic normal distributions,

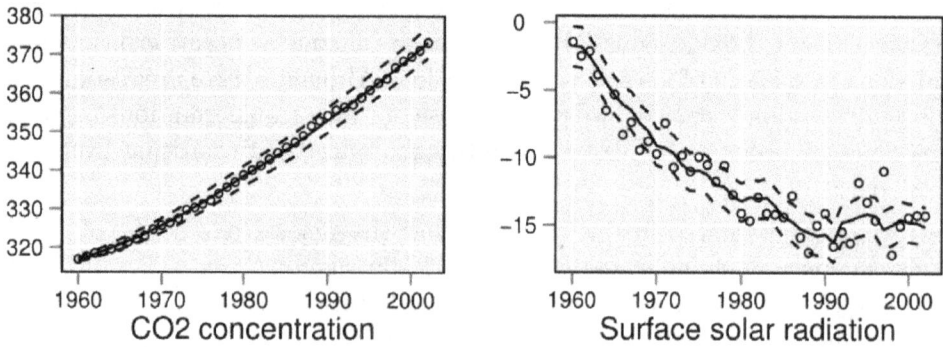

Figure 2: In-sample predictions for CO2 concentration, measured in parts per million by volume (ppmv) (left panel), and surface solar radiation, measured in Watts per meter squared (Wm^{-2}), in deviation of the average level of surface solar radiation in 1959 (right panel).

representing parameter uncertainty, while the error terms are drawn from the corresponding sample error distributions, representing process uncertainty. The dashed curves depict the range of 'likely outcomes', based on these 5000 simulations. (In contrast to statistical practice, where 95% bands are typically used, the IPCC defines an outcome as 'likely' when it falls within the 67% confidence band.) We conclude that our models describe the data with sufficient accuracy.

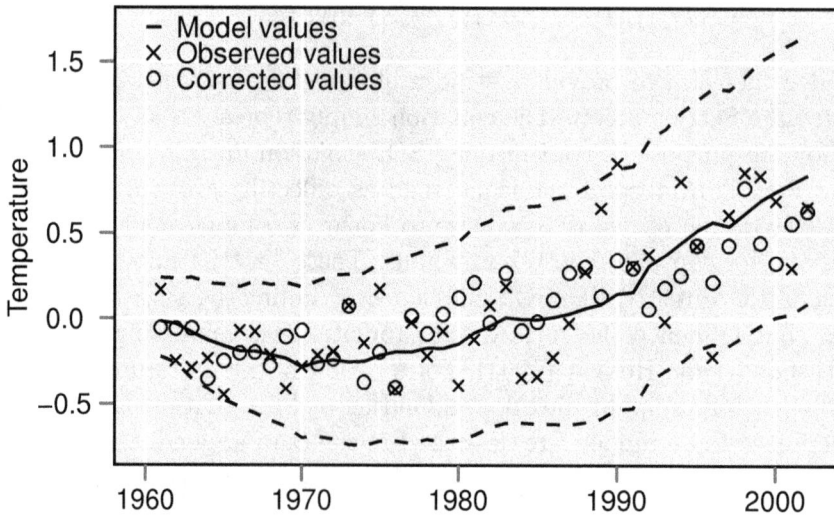

Figure 3: In-sample predictions for temperature, measured in degrees Celsius (°C), with the temperature in 2002 set equal to 0.8 °C.

Further evidence is provided by Figure 3, which shows that the trends in the temperature series are well captured by our simple model. The points denoted × are the observed values and the small circles are the values corrected for measurement error. The solid line gives the in-sample fit and the dashed lines are 67% confidence bands. All curves are normalized so that their in-sample mean coincides with the mean of the temperature series. The temperature is measured with the pre-industrialization temperature set equal to 0 °C. On average, the corrected measurements are closer to the model values, and both the observed and corrected values are (broadly) within the 67% confidence band.

6. Scenarios

We now have an integrated model of climatic change, estimated its parameters, and verified that the model provides a satisfactory description of the temperature series. Before we can analyze scenarios in the next section, we need to decide which scenarios we wish to consider.

We shall consider three scenarios. First, a baseline model, denoted '00', in which carbon dioxide and sulphur dioxide emissions are kept constant at their end-of-sample values (2002). The other two scenarios are publicly available IPCC scenarios (SRES) to facilitate comparison between our projections and those of other modeling groups. The SRES scenarios that we analyze are known as 'A1T' and 'A1FI'.

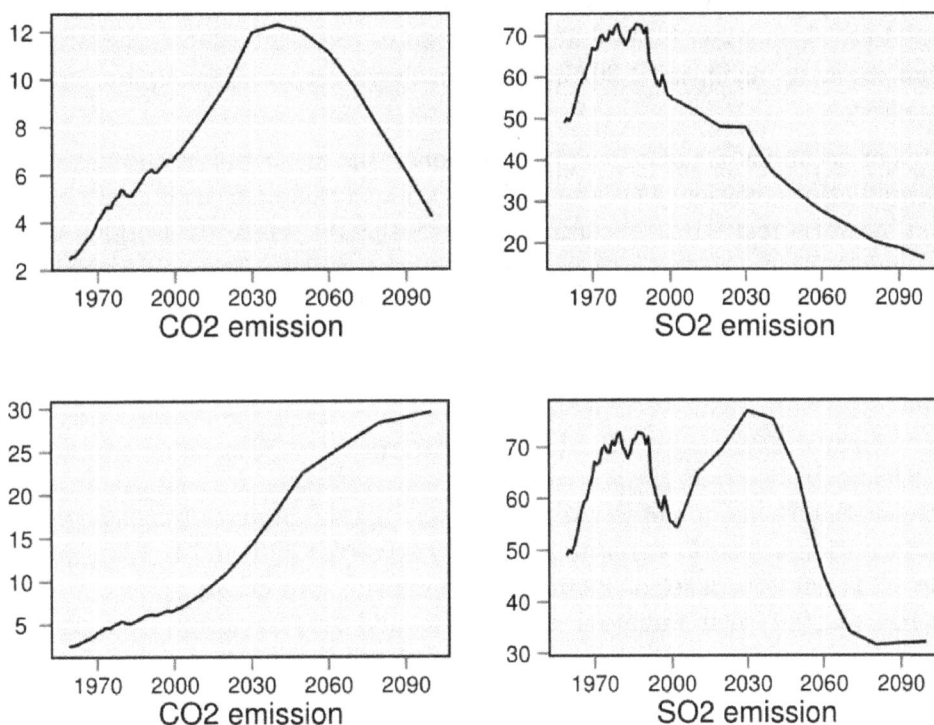

Figure 4: Emissions scenarios (top: A1T, bottom: A1FI) for CO2, measured in millions of metric tons of carbon (left panel), and SO2, measured in metric tons of sulfur (right panel). [Caption goes here]

Both scenarios are part of the A1 storyline, which postulates rapid and continuing economic growth, a global population that reaches around 9 billion in 2050 and then gradually declines, and a rapid introduction of new and more efficient technologies. The main difference between

A1T and A1FI is that A1T emphasizes the use of alternatives to fossil energy sources, while A1FI describes a world that intensively uses fossil energy sources. For more information about the IPCC/SRES scenarios, see Nakicenovic and Swart (2000).

The SRES SO2 emission scenarios do not match the realization in the period 1990–2000. A reversal of the trend after 1990 was already reported by Stern (2006). In particular, the economic downturn in the former USSR and Eastern Europe after 1989 resulted in a substantial reduction in SO2 emissions, and the SRES scenarios do not take this reduction into account. If we were to use the post-2000 levels of the SRES scenarios without adaptation, then a serious jump-off bias would result, compared to the actual in-sample values. For this reason, we create adapted SRES scenarios which do not suffer from jump-off bias. We follow the same strategy for the CO2 emission scenarios, although here the jump-off bias is much less serious.

To illustrate the resulting emission scenarios, Figure 4 presents a time- series plot of the carbon dioxide and sulphur dioxide emissions corresponding to scenarios A1T and A1FI. Our resulting scenarios are similar to the original SRES scenarios, but adapted to the actual levels. The emissions are quite different in the two scenarios. The carbon dioxide emission reaches a maximum around 2040 in scenario A1T, but continues to increase in scenario A1FI during the whole period, be it with some slowdown in growth after 2050. The sulphur dioxide emission continues to decrease in scenario A1T after the drop in the 1990s, but in scenario A1FI the emission first increases again until 2040, and then decreases to become more or less stable from 2080 onwards.

To complete our scenarios, we also have to extrapolate the linear time trend in the empirical surface solar radiation equation (17). Our interpretation of this time trend is economic growth as measured by world real GDP. According to the SRES A1 scenarios, world real GDP will continue to increase more or less exponentially, so that its logarithm will increase more or less linearly as a function of time. Thus we extrapolate the linear time trend in the empirical surface solar radiation equation linearly by extending the linear in-sample trend.

7. Projections

We now present the results of our scenario analysis. Figure 5 projects the changes in carbon dioxide concentration and surface solar radiation for scenarios A1T and A1FI. More precisely, for scenario A1T, the top panel of Figure 5 presents the projections of the estimated carbon and aerosol-surface solar radiation models (16) and (17), resulting from the emissions illustrated in the top of Figure 4, together with the time trend in the empirical surface solar radiation equation, extrapolated linearly. The bottom panel of Figure 5 provides the projections for scenario A1FI, based on the bottom panel of Figure 4. The dashed lines correspond to 67%-confidence intervals, taking into account both parameter and process uncertainty.

The uncertainty is more pronounced as we move forward in time, exactly as one would expect. The conversion from CO2 emissions to CO2 concentration is slow, which is why the resulting graphs are smoother than their inputs. The decrease in CO2 emission after 2040 in the A1T scenario results in a slowdown of CO2 concentration growth. In the A1FI scenario the growth in CO2 concentration speeds up over time. The surface solar radiation patterns follow the SO2 patterns closely, but with a downward trend over time due to the extrapolated linear time trend, reflecting continuous economic growth as measured by GDP (see also section 5,

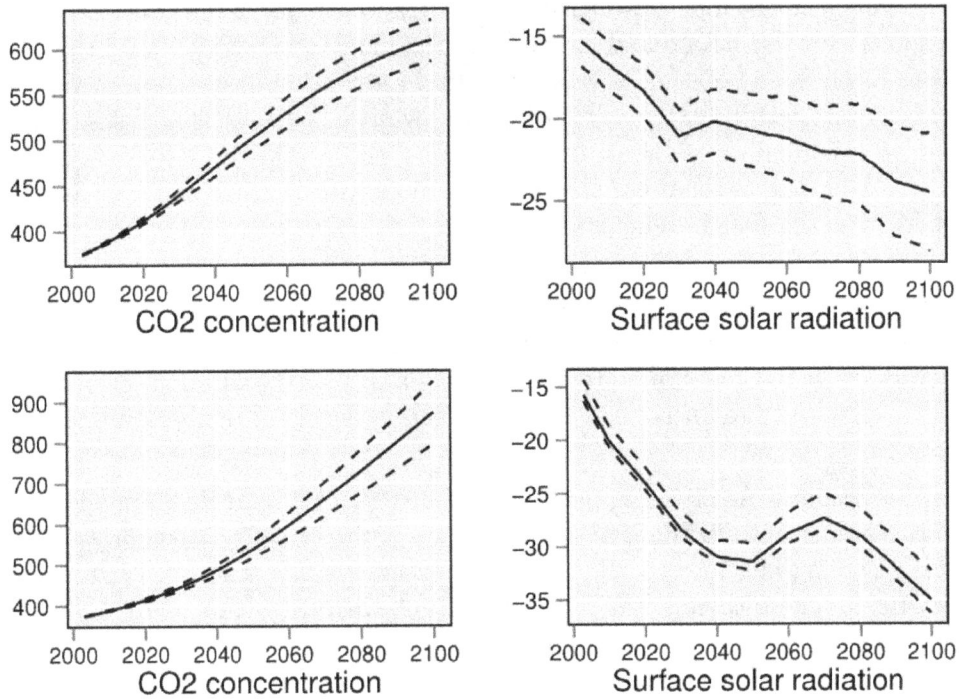

Figure 5: Projections (top: A1T, bottom: A1FI): CO2 concentration, measured in parts per million by volume (ppmv) (left panel), and surface solar radiation, measured in Watts per meter squared (Wm^{-2}), in deviation of the average level of surface solar radiation in 1959 (right panel).

where we link economic growth to emissions not captured by SO2 emissions).

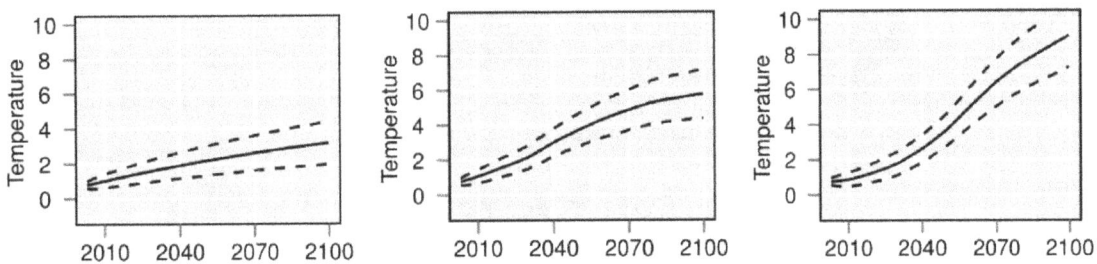

Figure 6: Temperature projections: 00 (left), A1T (middle), and A1FI (right). Temperature is measured in degrees Celsius (°C), with the temperature in 2002 set equal to 0.8 °C.

Figure 6 presents our probabilistic temperature projections for the three scenarios 00, A1T, and A1FI. The solid curve shows the temperature forecast (mean over 5000 simulations), while the dashed curves represent the range of likely outcomes (the 67%-confidence interval), accounting for both parameter and process uncertainty. The temperature is measured with the pre-industrialization temperature set to zero.

Without any change in the inputs (scenario 00, left panel), the mean temperature in 2100 (averaged over all scenarios) lies around 3.1 °C, and the likely range is $(1.9°C, 4.4°C)$. Keeping the emissions at their end-of-sample values results in a continuing increase of CO2 concentration, which has a positive effect on temperature, while a constant SO2 emission level does not result in an extra reduction (apart from the time trend) in average surface solar radiation according to our empirical relation (17), and thus no extra negative effect on temperature. The net effect is an increase in temperature.

In scenario A1T (middle) we project a temperature increase in 2100 of about 5.7°C, with a likely range of $(4.3°C, 7.0°C)$. CO2 emissions are higher than in scenario 00, resulting in higher CO2 concentration and higher temperature. At the same time, the decreasing SO2 emission levels result in only moderate changes in surface solar radiation, corresponding to only moderate changes in temperature. The net effect is a stronger temperature increase than in scenario 00.

The results for scenario A1FI (right panel) are based on CO2 and SO2 emissions that are higher than in scenario A1T. Higher CO2 emissions positively affect temperature, and this detoriates the cooling effect of the increase in SO2 emissions. The resulting temperature increase in 2100 is higher than for A1T, around 8.9°C, with likely range $(7.2°C, 10.7°C)$.

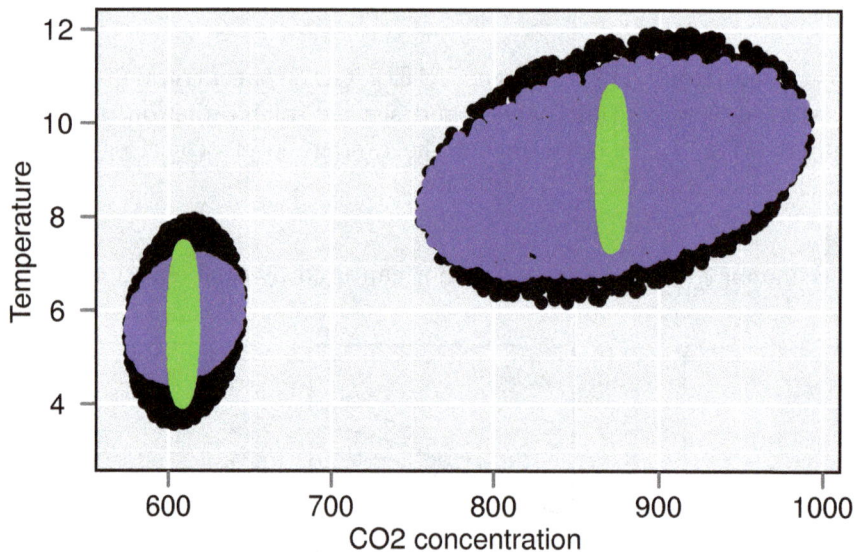

Figure 7: Projections of likely temperature and CO2 concentration combinations (left: A1T, right: A1FI). Temperature is measured in degrees Celsius (°C), with the temperature in 2002 set equal to 0.8 °C. CO2 concentration is measured in parts per million by volume (ppmv).

In Figure 7 we complement the projections of Figure 6 by presenting the joint 67%-confidence set for CO2 concentration and temperature. We distinguish between process risk (green), parameter risk (blue), and combined risk (red). The lower-left confidence set corresponds to scenario A1T, and the upper-right confidence set to scenario A1FI. The process risk (green areas) is more or less of the same size in both cases. But the higher input levels in scenario A1FI generate substantially more parameter risk than in scenario A1T, resulting in a much larger 67%-confidence set. This is particularly so in the CO2 concentration dimension.

Compared to Solomon et al. (2007, Chapter 10) we find similar levels in CO2 concentrations, but higher temperature increases in both scenarios A1T and A1FI. Focusing on temperature, we do find similar lengths of the likely ranges. For scenario A1T, Solomon et al. (2007) report a warming of 2.4°C with a likely range of 2.4°C (following from the interval (1.4°C, 3.8°C)), versus a likely range of 2.7°C in our case, and for scenario A1FI they report a warming of 4.0°C with a likely range of 4.2°C (following from the interval (2.4°C, 6.4°C)), versus a likely range of 3.5°C in our case. Thus, the degree of uncertainty in our projections is comparable to Solomon et al. (2007). The IPCC presents indirectly validated multi-model outcomes, whereas our outcomes can be seen as a *direct* outcome based on statistical confidence sets, given our statistical model. This suggests that from our statistical perspective sufficient uncertainty is incorporated in the IPCC scenarios. In this sense, our results validate the IPCC outcomes.

However, we find higher temperature projections than the IPCC, although the results are not directly comparable, because Solomon et al. (2007) report the forecasted temperature change during the ten years 2090–2099 relative to 1980–1999, while we report the end-of-period (that is, in the year 2100) temperature relative to the pre-industrialization temperature level. But even when we correct for these differences, our projected temperature levels remain substantially higher. Perhaps the reason is that our simple statistical model is too simple, by ignoring important forces, such as self-regulatory mechanisms in the climate system, which are incorporated into the more advanced models employed by the IPCC. On the other hand, our more alarming empirical findings are not without support. A study by Rahmstorf et al. (2007), for example, indicates that (up to 2006) an aerosol cooling smaller than expected might be a possible cause of a realized warming in the upper part of the range projected by the IPCC. In our study we do incorporate lower levels of SO2 emissions, yielding less aerosol cooling, in line with these empirical findings.

Our temperature projections are thus located on the edge of the IPCC range; they are not in conflict with the IPCC outcomes. For scenario A1T our 95%-confidence interval (for the year 2100, relative to the pre-industrialization temperature level) is given by (3.4°C, 8.1°C), and in scenario A1FI it is given by (5.5°C, 12.8°C). Thus, using 95%-confidence intervals (and correcting for the differences in reporting), there is substantial overlap with the results by Solomon et al. (2007).

8. Conclusions

In this paper a simple model of climate change was presented —simple enough to allow rigorous statistical analysis. The analysis consisted of quantifying the uncertainty associated with projections of future climate change. We introduced our model, presented the parameter estimates, and showed that our simple model describes historical climate change well. Then we used the model to generate predictions of future climate change and, most importantly, we quantified the uncertainty associated with these predictions, distinguishing between process and parameter risk.

For the scenarios considered (the SRES scenarios A1T and A1FI), our model predicts an increase in temperature above the best guess in the most recent IPCC report (Solomon et al., 2007). However, given the range of uncertainty around our projection (quantified by

95%-confidence intervals), the IPCC projections are not ruled out. We also find, in our single model, that the uncertainty range due to parameter and process uncertainty is of the same order as the uncertainty range reported in Solomon et al.'s (2007) multi-model projections, with the empirically-based parameter risk being the dominant source of risk. One application of our statistical analysis is therefore to serve as empirical validation of the multi-model approach employed by the IPCC, which only allows for an indirect validation. Another application would be to obtain a quick, first impression of climate change consequences, including the corresponding uncertainty, under alternative scenarios.

Acknowledgments

We are grateful to Reto Knutti and an anonymous Associate Editor for helpful comments and discussion.

Appendix: Emission data

We use global data on carbon dioxide emission and sulphur dioxide emission. CO_2 emissions are measured in millions of metric tons of carbon, while SO_2 emissions are measured in metric tons of sulfur. For CO_2 emissions we use data over the years 1959–2002 from the Carbon Dioxide Information and Analysis Center (CDIAC), available at the website

http://cdiac.ornl.gov/aboutcdiac.html.

For SO_2 emissions we use data over the years 1959–2000 from David Stern's website

http://www.sterndavidi.com/datasite.html.

Data over the years 2001 and 2002 are lacking. For this reason, we estimated and applied (17) twice: once using only the available data 1959–2000, and once using these data extended by assuming that our SO_2 emission data in 2001 and 2002 have the same growth as the growth of the "grand total" SO_2 emissions of EDGAR v4.1. Both outcomes are quite similar. In the main text we use and report the second case only. The source of the EDGAR v4.1 data is: European Commission, Joint Research Centre (JRC)/Netherlands Environmental Assessment Agency (PBL). Emission Database for Global Atmospheric Research (EDGAR), release version 4.1, with corresponding website

http://edgar.jrc.ec.europa.eu, 2010.

References

Andreae MO, Jones CD, Cox PM (2005). "Strong Present-Day Aerosol Cooling Implies a Hot Future." *Nature*, **435**, 1187–1190.

Arellano M, Bond SR (1991). "Some Tests of Specification for Panel Data: Monte Carlo Evidence and an Application To Employment Equations." *Review of Economic Studies*, **58**, 277–297.

Blundell R, Bond SR (1998). "Initial Conditions and Moment Restrictions in Dynamic Panel Data Models." *Journal of Econometrics*, **87**, 115–143.

Brohan P, Kennedy JJ, Harris I, Tett SFB, Jones PD (2006). "Uncertainty Estimates in Regional And Global Observed Temperature Changes: A New Dataset From 1850." *Journal of Geophysical Research*, **111**, D12106.

Gilgen H, Ohmura A (1999). "The Global Energy Balance Archive." *Bulletin of the American Meteorological Society*, **80**, 831–850.

Härdle W, Linton O (1994). "Applied Nonparametric Methods." In D McFadden and RF Engle (eds.), "The Handbook of Econometrics, Vol. 4," New Holland, New York, 2295–2339.

Knutti R, Hegerl GC (2008). "The Equilibrium Sensitivity of the Earth's Temperature to Radiation Changes." *National Geoscience*, **1**, 735–743.

Knutti R, Plattner GK (2012). "Comments on 'Why Hasn't Earth Warmed as Much as Expected?'." *Journal of Climate*, **25**, 2192–2199.

Magnus JR, Melenberg B, Muris C (2011). "Global Warming and Local Dimming: The Statistical Evidence." *Journal of the American Statistical Association*, **106**, 452–464.

Majda A, Timofeyev I, Eijnden E (2001). "A Mathematical Framework for Stochastic Climate Models." *Communications on Pure and Applied Mathematics*, **54**, 891–974.

Meehl GA, Stocker TF, Collins WD, Friedlingstein P, Gaye AT, Gregory JM, Kitoh A, Knutti R, Murphy JM, Noda A, Raper SCB, Watterson IG, Weaver AJ, Zhao ZC (2007). "Global Climate Projections." In: S Solomon, D Qin, M Manning, Z Chen, M Marquis, KB Averyt, M Tignor, HL Miller (eds.), "Climate Change 2007: The Physical Science Basis. Contribution of Working Group I to the Fourth Assessment Report of the Intergovernmental Panel on Climate Change," Cambridge University Press, Cambridge, UK.

Mitchell TD, Jones PD (2005). "An Improved Method of Constructing a Database of Monthly Climate Observations and Associated High-Resolution Grids." *International Journal of Climatology*, **25**, 693–712.

Nakicenovic N, Swart R (eds.) (2000). *Emissions Scenarios*. Cambridge University Press, Cambridge, UK.

Padilla LE, Vallis GK, Rowley CW (2011). "Probabilistic Estimates of Transient Climate Sensitivity Subject to Uncertainty in Forcing and Natural Variability." *Journal of Climate*, **24**, 5521–5537.

Rahmstorf S, Cazenave A, Church JA, Hansen JE, Keeling RF, Parker DE, Somerville RCJ (2007). "Recent Climate Observations Compared to Projections." *Science*, **316**, 709.

Roe GH, Baker MB (2007). "Why is Climate Sensitivity So Unpredictable?" *Science*, **318**, 629–632.

Schwartz SE, Charlson RJ, Kahn RA, Ogren JA, Rodhe H (2010). "Why Hasn't Earth Warmed as Much as Expected?" *Journal of Climate*, **23**, 2453–2464.

Solomon S, Qin D, Manning M, Chen Z, Marquis M, Averyt KB, Tignor M, Miller HL (eds.) (2007). *Contribution of Working Group I to the Fourth Assessment Report of the Intergovernmental Panel on Climate Change*. Cambridge University Press, Cambridge, UK.

Stainforth DA, Aina T, Christensen C, Collins M, Faull N, Frame DJ, Kettleborough JA, Knight S, Martin A, Murphy JM, Piani C, Sexton D, Smith LA, Spicer RA, Thorpe AJ, Allen MR (2005). "UncertainTy in Predictions of the Climate Response to Rising Levels of Greenhouse Gases," *Nature*, **433**, 403–406.

Stern DI (2006). "Reversal of the Trend in Global Anthropogenic Sulfur Emissions." *Global Environmental Change*, **16**, 207–220.

Tselioudis G, Rossow WB (1994). "Global, Multiyear Variations of Optical Thickness with Temperature in Low and Cirrus Clouds." *Geophysical Research Letters*, **21**, 2211–2214.

Vallis GK, Gerber EP, Kushner PJ, Cash BA (2004). "A Mechanism and Simple Dynamical Model of the North Atlantic Oscillation and Annular Modes." *Journal of the Atmospheric Sciences*, **61**, 264–280.

Wild M. (2009). "GlobAl Dimming and Brightening: A Review." *Journal of Geophysical Research*, **114**, D00D16.

Wild M, Grieser J, Schär C (2008). "Combined Surface Solar Brightening and Increasing Greenhouse Effect Support Recent Intensification of the Global Land-Based Hydrological Cycle." *Geophysical Research Letters*, **35**, L17706.

Wild M, Ohmura A, Gilgen H, Rosenfeld D (2004). "On the Consistency of Trends in Radiation and Temperature Records and Implications for the Global Hydrological Cycle." *Geophysical Research Letters*, **31**, L11201.

Affiliation:

Jan R. Magnus
Department of Econometrics and Operations Research
Vrije Universiteit Amsterdam
De Boelelaan 1105
1081 HV Amsterdam, The Netherlands
Email: jan@janmagnus.nl

Bertrand Melenberg
Department of Econometrics and Operations Research
Tilburg University
PO Box 90153
5000 LE Tilburg, The Netherlands
E-mail: b.melenberg@uvt.nl

Chris Muris
Department of Economics
Simon Fraser University
8888 University Drive
Burnaby, BC V5A 1S6, Canada
Email: cmuris@sfu.ca

Martin Wild
ETH Zürich
Institute for Atmospheric and Climate Science
CHN L 16.2
Universitätstrasse 16
8092 Zürich, Switzerland
Email: martin.wild@env.ethz.ch

9

Temporal Investigation of Flow Variability in Scottish Rivers Using Wavelet Analysis

Maria Franco-Villoria, Marian Scott
School of Mathematics and Statistics, University of Glasgow

Trevor Hoey
School of Geographical
and Earth Sciences, University of Glasgow

Denis Fischbacher-Smith
Business School
University of Glasgow

Abstract

River flow records form the basis of flood risk estimates. In Scotland, the Flood Risk Management Act (2009) has the objective of improving flood risk management by improved modelling of river flows taking into account potential future climate change. Wavelet analysis is presented here as a possible method for detecting and comparing trends, investigating spatial heterogeneity and periods of significant variability in non-stationary environmental time series. The results from a wavelet analysis of a set of 9 Scottish rivers confirm a difference in river flow maxima between the East and the West that has been pointed out in previous studies, along with changes in the seasonal patterns that may be linked to external climatic drivers, including the North Atlantic Oscillation and the Atlantic Multidecadal Oscillation. Such influences (which act on several time scales, the principal one being annual) are not constant and vary both temporally (being stronger from 1987 onwards) and spatially, with a possible catchment size effect, highlighting the importance of assessing flood risk at a regional level. The study is currently being extended to a further 26 rivers to gain a better understanding of the spatial dependence in extreme river flows, which will contribute to the improvement of flood risk management.

Keywords: discrete wavelet transform, stationarity, variability, river flow data.

---

1. Introduction

Understanding the pattern of flows and its relationship to flooding is critical to flood planning and risk management. In particular in Scotland the Flood Risk Management Act (2009) was passed with the aim of introducing "... a more sustainable and modern approach to flood risk management" (Government (2010)). To do so, new and improved estimates of flood risk which take into account the impact of climate change and possible spatial heterogeneity are needed.

Records of river flows are widely used to predict flood and low flow levels, in water resource allocation, and form an important basis for assessing the impacts of climate change. Data records are often short (a few decades or less) and a range of classical time-series and extreme value methods have been used as a basis for making predictions. Much flood-risk management is based on the concept of a return period for an event of given magnitude and, despite ongoing debate about the utility of the return period approach (White (2001); Young and Davies (1989)), considerable effort continues to be made to refine predictions of the 1 in 100 year event. Current changes in environmental conditions, specifically those driven by climate change, have led investigators to look for new statistical models which account for non-stationarity and seasonal variations. Climate change impacts are also expected to vary spatially (Jenkins et al. (2009)) and thus would be expected to result in spatially variable changes in river flows. In addition, there is continuing evidence for spatial relationships, e.g. in Scotland, evidence of an East West difference in terms of rainfall and river flow is apparent over the last 30 years (Black (1996); Black and Burns (2002); Werritty (2002); Mayes (1996)). The attention drawn to this topic over the last few years coupled with the greater availability of data as a result of the recent development of climate models has led researchers to try to relate patterns in rainfall and river flow to various climate signals. In Europe, the main influence comes from the North Atlantic Ocean, for which two main indices have been identified, the North Atlantic Oscillation (NAO) and the Atlantic Multidecadal Oscillation (AMO).

The NAO is a large-scale signal of natural climate variability, calculated as the normalized atmospheric pressure difference (at sea level) between the Azores and Iceland (Hurrell (1995)). The resulting index is positive when the pressure is high in the Azores and low in Iceland, and negative when the situation is reversed. A high index is associated with strong westerly winds, cool summers, mild winters and frequent rain in the north of Europe, while a low index is linked to scarce winds, extreme temperatures and dry conditions (with localized storms) (Macklin and Rumsby (2007); Bouwer et al. (2008)).

A less well known and less studied index is the Atlantic Multidecadal Oscillation (AMO) (Kerr (2000); Delworth and Mann (2000)), a further signal of climatic variability, related to anomalies in the sea surface temperature (SST) and sea level pressure (Delworth and Mann (2000)) and calculated based on the former. Similarly to the NAO, the resulting index can be either positive (warm phase) or negative (cold phase). It has been attributed with an oscillation period of about 60 years and it is thought to be correlated to air temperatures and rainfall over much of the Norther Hemisphere, in particular affecting European summer climate (Knight et al. (2006); Sutton and Hodson (2005)).

Time series analysis can be approached from two different perspectives: the time domain (classical time series analysis) or the frequency domain. In hydrology, Fourier analysis (also called spectral analysis) has traditionally been used for the latter. However, it assumes that the time series is stationary and that it can be expressed as a "... linear superposition of linear, independent and non-evolving cycles" (Labat (2005)), conditions rarely met by hydrologic series. Having found clear evidence of non-stationarity by means of traditional time series analysis, newer statistical methodology, namely wavelets are presented here and applied to river flow data to investigate changes in river flow maxima patterns and evidence of spatial differences. Wavelet analysis is a useful tool for analysing non-stationary time series to capture the local behaviour at different frequencies (Percival and Walden (2006); Torrence and Compo (1998); Labat (2005)). By subsequently filtering the original series, we obtain sequences of results which relate to variations at different scales (frequencies). The result is a time-scale decomposition of the original time series that provides an alternative way of looking at the time series, showing features that would not be visible in, say, a plot of the time series versus time (Percival and Walden (2006); Torrence and Compo (1998)). All the information contained in the original time series is also preserved in its wavelet transform. Examples of wavelet analysis applied to hydrologic series can be found in Smith *et al.* (1998); Percival and Mofjeld (1997); Labat *et al.* (2005); Rossi *et al.* (2009) and Sen (2009).

In this paper, the trend and seasonal components of flows in 9 Scottish rivers are identified by means of wavelet analysis and compared for Western and Eastern rivers, and the influence that external climatic drivers such as the AMO and the NAO might have had on them is investigated. All the analysis was carried out using the software packages **sowas** and **wmtsa** in R (version 2.8.1).

1.1. Data

Given the country's geographical situation, Scotland's weather is subject to both continental and Artic influences, combined with a dominant Atlantic signal. There is also a strong West-East rainfall gradient; in particular, three precipitation regions have been defined, the South West, the North West and the East (Gregory *et al.* (1991)), with the North West being the wettest region and the East the driest. Apart from these three broad groups, local conditions vary from catchment to catchment within the same precipitation region. Nine rivers of different catchment sizes across Scotland (Figure 1) were selected based on data quality and spatial location. Since the main interest lies in the extreme values the (logged) series of monthly maxima were calculated. The resulting time series for one of the rivers is shown in Figure 2. Data (gauged daily flow (m^3/s)) were provided by the National River Flow Archive (NRFA) and the Scottish Environment Protection Agency (SEPA) (Table 1).

The NAO data were downloaded from *http://www.cdc.noaa.gov/data/climateindices/List/* and consists of a monthly series covering 51 years (January 1950-December 2008). Data for the AMO were downloaded from *http://www.esrl.noaa.gov/psd/data/timeseries/AMO/* and consists of a monthly series running from January 1856 to April 2009.

2. Methods - wavelet analysis

One way of identifying the local behaviour of non-stationary time series is by wavelet analysis. By subsequently filtering the original series, we obtain sequences of results which relate to variations at different scales (frequencies). The result is a time-frequency representation of the data (Percival and Walden (2006); Torrence and Compo (1998)). The discrete wavelet

Table 1: Data set.

River	Catchment area (km^2)	Mean discharge (m^3/s)	Location	Coordinates	Data record
Nith(Friars Carse)	799	27.38	West	3° 41.4' W,55° 08.9' N	1/10/57 - 1/10/08
Dee(Woodend)	1370	36.95	East	2° 36.1' W,57° 03.0' N	1/10/29 - 30/9/08
Tweed(Norham)	4390	79.59	East	2° 09.7' W,55° 43.4' N	1/10/62 - 31/10/08
Ewe(Poolewe)	441	29.91	West	5° 36.0' W,57° 45.7' N	19/10/70 - 31/12/08
Lossie(Sheriffmills)	216	2.70	East	3° 21.0' W,57° 38.8' N	1/10/63 - 31/10/07
Tay(Ballathie)	4857	169.22	East	3° 23.7' W,56° 45.4' N	1/10/52 - 31/10/08
Water of Leith(Murrayfield)	107	1.49	East	3° 14.2' W,55° 56.7' N	1/1/63 - 31/12/05
Clyde(Blairston)	1704	42.80	West	4° 04.1' W,55° 47.8' N	1/10/58 - 5/11/08
Kelvin(Killermont)	335	8.41	West	4° 18.4' W, 55° 54.4' N	1/10/48 - 31/12/07

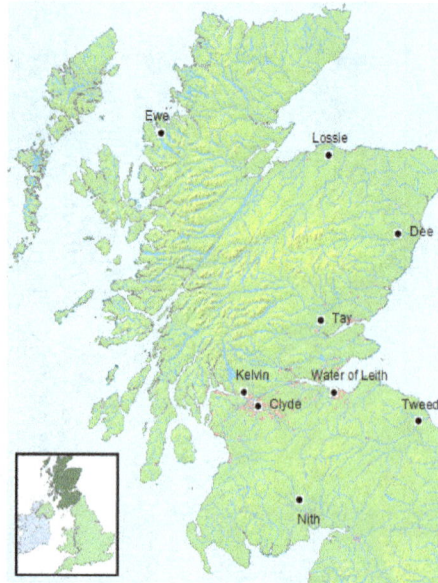

Figure 1: Map of Scotland showing the gauging stations used in the analysis. Source: *map-soft.net*

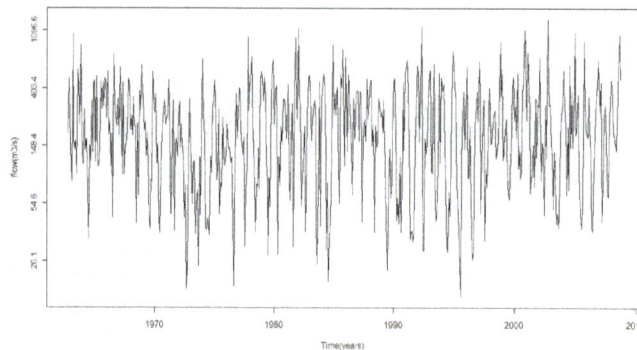

Figure 2: Time series of logged monthly maxima (River Tweed).

transform (DWT) coefficients $\{W_n : n = 1,\ldots,N\}$ of a time series X of length N are calculated as $\mathbf{W} = \mathcal{F}\mathbf{X}$, where $\mathcal{F}$ is an N×N matrix constructed using the chosen filter. The original time series can then be reconstructed as the sum of a number of wavelet detail components D_j and a smooth component S_J :

$$X = \mathcal{F}^T \mathbf{W} = \sum_{j=1}^{J} D_j + S_J \qquad (1)$$

where J is the level of decomposition. D_j (wavelet detail) is a time series related to variations in X at scale $\tau_j = 2^{j-1}$, j=1,\ldots,J, and S_J (wavelet smooth) is a time series associated with scales $\lambda_J = 2^J$ and higher and can be interpreted as the trend. The DWT has a number of constraints; first, the sample size of the time series has to be a power of two, and second, both the filter choice and the starting point of the time series might have an influence on the resulting wavelet transform. These constraints are due to the algorithm underlying the construction of the discrete wavelet transform, in which the filtered series is downsampled at each stage (Percival and Walden (2006)). The maximum overlap discrete wavelet transform (MODWT) provides a decomposition of the time series without the constraints of the DWT. The algorithm of the DWT is modified so that no downsampling is involved; as a result, it is no longer an orthogonal transformation and the computational cost is higher (Percival and Walden (2006)). Similarly to the DWT, the original series can be expressed as the sum of a number of wavelet details components plus a smooth component. By investigating the wavelet decomposition of a time series, one can identify the scale(s) responsible for the major part of the variability. The time-dependent wavelet variability at a particular scale τ_j ($\hat{v}_{X,t}(\tau_j)$) describes the behaviour of the variability at that scale over time, and can be estimated as:

$$\hat{v}_{X,t}(\tau_j) = \frac{1}{N_S} \sum_{u=-(N_S-1)/2}^{(N_S-1)/2} W_{j,t+|\nu_j^{(H)}|+u \, mod \, N} \tag{2}$$

where $u \, mod \, N$ is the remainder of u/N, N_S is the width of the smoothing window and $W_{j,t+|\nu_j^{(H)}|+u \, mod \, N}$ is the wavelet coefficient $W_{j,t}$ circularly shifted so that it is aligned in time with the original time series (Percival and Walden (2006)). In particular, we chose $N_S=12$ given the monthly resolution of the data.

A continuous version of the wavelet transform is also available. When the interest is to study how two time series are related, continuous wavelet analysis offers a wavelet-based version of the usual cross-correlation, to investigate the relationship between two time series not only across time but also for different time scales. The cross-wavelet spectrum of two time series X and Y is defined as:

$$W_n^{XY}(s) = W_n^X(s)W_n^{Y*}(s) \tag{3}$$

where $W_n(s)$ is the continuous wavelet transform (Torrence and Compo (1998)) at time n and scale s and * indicates the complex conjugate. However, defined this way it is not a reliable tool, for peaks of high correlation can appear even when the two series are independent, reflecting just peaks of high variability of the individual series (Maraun and Kurths (2004)). The alternative is to use a normalized version of the cross-wavelet spectrum, the wavelet coherency. The wavelet cross spectrum is a complex number, so it can be re-written as $W_n^{XY}(s) = |W_n^{XY}(s)|e^{n\phi_n(s)}$. The wavelet coherency (Grinsted et $al.$ (2004); Torrence and Webster (1999)) is then defined as:

$$WCO_n^{XY}(s) = \frac{|\langle s^{-1}W_n^{XY}(s)\rangle|^2}{\langle s^{-1}|W_n^X(s)|^2\rangle\langle s^{-1}|W_n^Y(s)|^2\rangle} \tag{4}$$

Its value ranges from 0 to 1. The symbol $\langle \cdot \rangle$ indicates smoothing. The smoothing, which can be done in time or/and scale direction, is necessary because otherwise the wavelet coherency would always be equal to 1, for every time point and scale (Torrence and Compo (1998)).

The wavelet coherency provides information about how strong the association between the two time series is, but it does not carry information about the time lag at which the two series are correlated. The phase function $\phi_n(s)$ provides a measure of the lag difference between the two time series at time n, scale s, and it can be calculated as:

$$\phi_n(s) = \tan^{-1}\left(\frac{Im\{\langle s^{-1}W_n^{XY}(s)\rangle\}}{Re\{\langle s^{-1}W_n^{XY}(s)\rangle\}}\right) \tag{5}$$

where $s^{-1}W_n^{XY}(s)$ has been smoothed.

A significance test for the wavelet coherency was proposed by Maraun and Kurths (2004) and Grinsted et $al.$ (2004), the null hypothesis being that the processes are not significantly correlated. However, the fact that neighboring times and scales are correlated is problematic for deriving the distribution of the test-statistic under the null hypothesis. To overcome this problem, Maraun and Kurths (2004) and Grinsted et $al.$ (2004) make use of Monte Carlo simulation to generate 10000 realizations of two independent Gaussian white noise processes or AR(1) processes. The simulated sample can be used to derive the empirical distribution of the test-statistic under H_0, from which a critical value for the chosen significance level can be obtained. Critical values depend on the amount of smoothing and appear to be scale dependent (Grinsted et $al.$ (2004)) even though Maraun and Kurths (2004) suggests that the ideal is to find the right amount of smoothing so that they are scale independent.

3. Results

3.1. Trends and seasonality

MODWT (based on an LA(8) filter (Figure 3) (Percival and Walden (2006)) with 4 levels of decomposition was applied to the monthly maxima series (normalized to the overall mean). LA (or least asymmetric) filters are a special class of filters proposed by Daubechies (1992). LA filters are appropriate for calculating the DWT because their phase function is very close to the phase function a linear phase filter. This means that the filtered series can be easily aligned in time with the original series (Percival and Walden (2006)). An alternative to Daubechies filters are Coiflet filters, but the use of the latter is not as common as they are likely to introduce artifacts in the wavelet transform (Percival and Walden (2006)). Since it is of interest to align in time the original series with its wavelet transform, an LA filter was considered appropriate. The number 8 here represents the width of the filter. Following Percival and Walden (2006), who suggest the width of the filter to be as small as possible, this value was chosen comparing a series of wavelet transforms calculated for a range of filter width values. The wavelet transform corresponding to filter width equal to 8 provided a good smooth representation of the corresponding time series; smaller width values resulted in sharp peaks in the individual elements of the time series decomposition, while greater width values did not make any difference. Each of the 9 river series was decomposed as $\sum_{j=1}^{4} D_j + S_4$. Components D_1, ..., D_4 reflect changes over the scales 1, 2, 4 and 8 months respectively. The smooth component S_4 reflects changes over a scale of 16 months (and higher) and can be regarded as the trend. Results are presented in Figures 4 and 5.

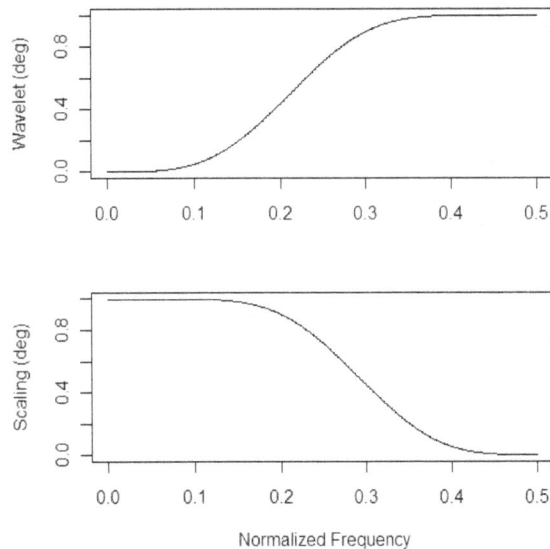

Figure 3: Wavelet and scaling filters for the LA(8).

Figure 4 confirms that the seasonal component varies over the years. Component D_3 is of special interest, as it reflects changes over a scale of 4 months and therefore can be regarded as the seasonal component. Also, D_3 is the main contributor to the sample variance for all 9 rivers.

The trend series (component S_4 from wavelet decomposition), plotted on Figure 5, suggest some differences between the East (Figure 5(a)) and the West (Figure 5(b)). There is a sharp decrease in the East around 1973 that, even though present in the West series, is not as low as for the latter. There is a second decrease around 1996 although in this case it is more prominent in the West than in the East. It is difficult to say whether there is an overall increasing or decreasing trend as the trend series are not linear; however, there appears to have been a slight increase in the West towards the end of the record that is not found in the East.

The time dependent wavelet variability (Equation 2) was calculated to get a measure of fluctuations in the yearly cycle variability. The resulting series are plotted on Figure 5. Overall, the variability is higher in the West (Figure 5(d)) than in the East (Figure 5(c)). There is a clear indication of non-stationarity with periods of high variability alternating with periods of very low variability. In particular, there is a clear change point around mid 1986 for both eastern and western rivers, when the variability is minimum.

3.2. Relationship with climatological covariates

Having found strong indication of changes in the seasonality of flow in these rivers, it is of

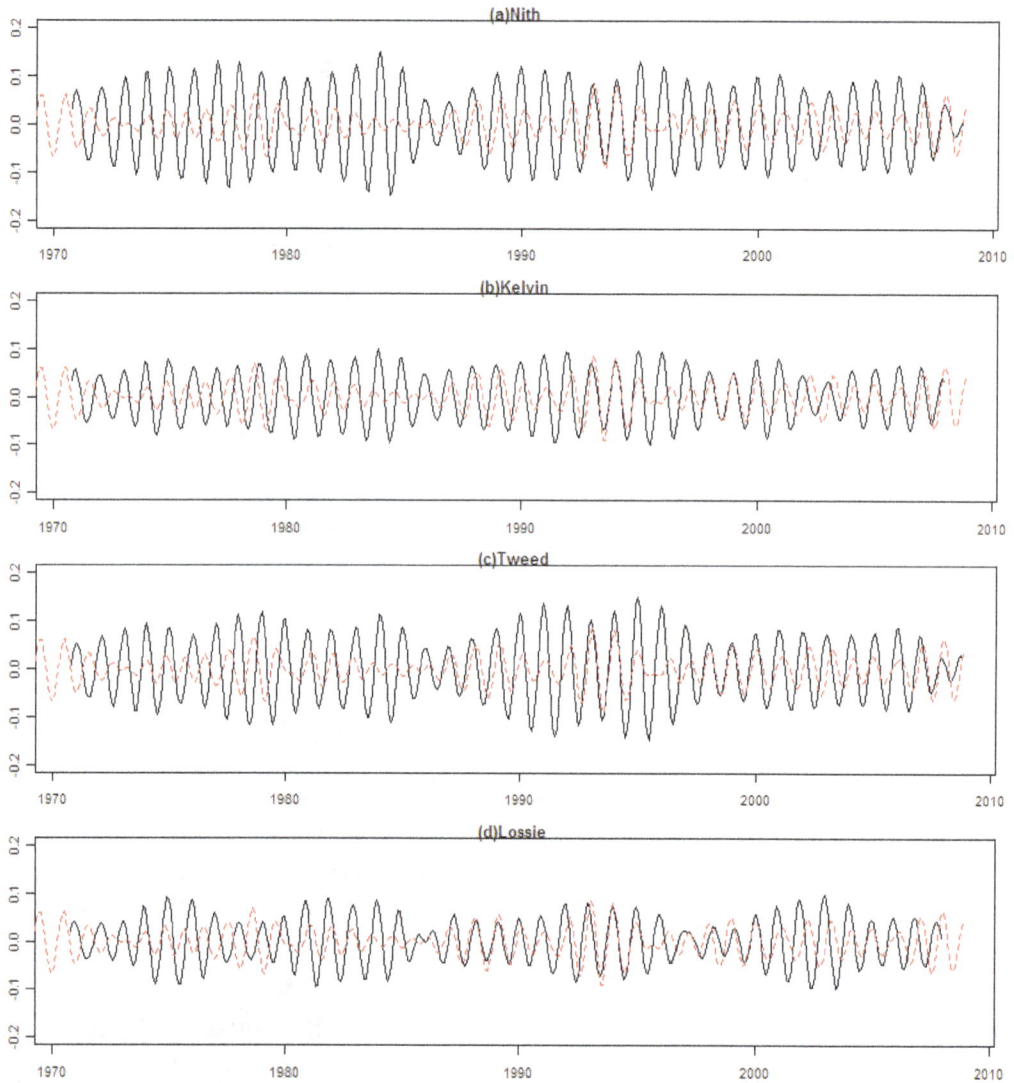

Figure 4: Seasonal component (D_3) for rivers (a)Nith, (b)Kelvin (West) and (c)Tweed and (d)Lossie (East). For reasons of space, only 4 rivers are plotted. The dashed red line represents the seasonal component for the NAO (see page 12).

interest to investigate the potential drivers of these changes as well as whether different regions of the country might be affected in different ways. Black (1996) compared Scotland with the rest of Europe, finding similar results in terms of increases in annual maximum flood in Norway, southern Finland and Estonia over the last 30-50 years, suggesting that there might be a common climate cause. The North Atlantic Oscillation (NAO) and the Atlantic Multidecadal Oscillation (AMO) are two potential causes of some of the changes.

The time series plot of the NAO series is shown in Figure 6(a), along with a long-term trend (S_4) estimated from wavelet analysis. The trend series shows that the NAO was mainly in

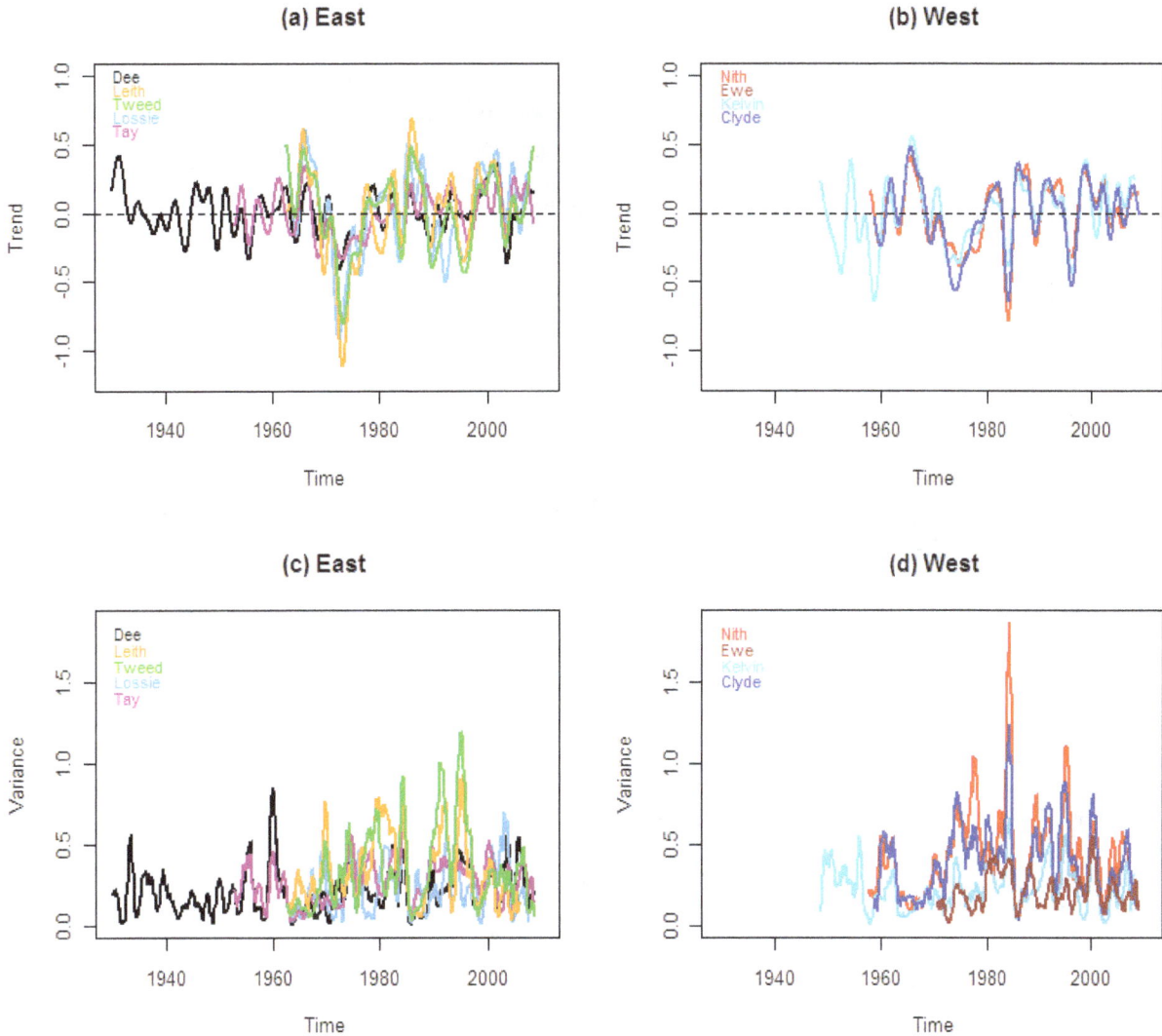

Figure 5: Trend (S_4) from wavelet decomposition for (a)Eastern and (b)Western rivers and seasonal time dependent variability based on component D_3 for (c)Eastern and (d)Western rivers.

negative phase up to the beginning of the 1970s and in positive phase during the 1980s and the first half of the 1990s.

Similarly, the AMO series is plotted in Figure 6(b), with the corresponding wavelet estimated long-term trend. Even though it has been less explored in the literature, it is of great interest to investigate how this index relates to river flow, and maxima in particular. The reason for this is that it is less variable than the NAO, with a fairly well established cyclic component and hence is more predictable. It is clear from the plot that the AMO was in its warm phase until the beginning of the 1960s and then again since the mid 1990s onwards, with a cold

(a) NAO

(b) AMO

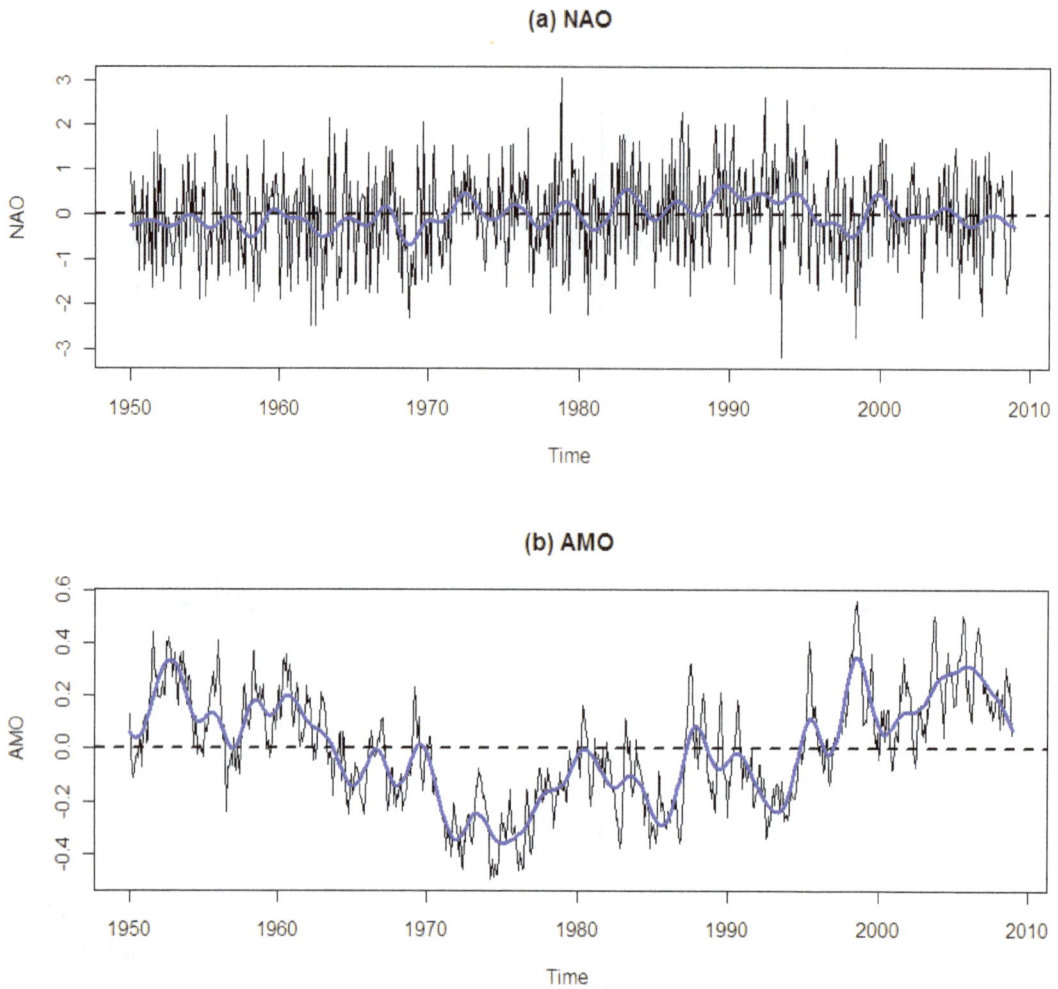

Figure   6:   (a)NAO   and   (b)AMO   time   series.   The   blue   line   is
the   trend   as   estimated   from   wavelet   analysis.   Data   were   downloaded   from   *http://www.cdc.noaa.gov/data/climateindices/List/*   and
http://www.esrl.noaa.gov/psd/data/timeseries/AMO/

phase of about 30 years in between. During this cold phase there seems to be a period just
before 1990 with anomalously high AMO values.

Even though some authors restrict the NAO influence to winter months, NAO has been shown
to be significantly correlated to precipitation over Northern Europe outside the winter months
(Marković and Koch (2005)). The results presented here are based on the whole NAO series. The length of the data record varies from river to river; for comparison purposes, only
values from October 1975 onwards have been used. The wavelet coherence (top) and phase
difference (bottom) (smoothed in time and scale direction) between NAO and each of the

rivers is shown on Figure 7. Only the graphs for 4 of the rivers are shown here for reasons of space. In general, the correlation takes values between 0.4 and 0.6 for most of the time and most scales. The thick black contour lines denote regions of statistically significant correlation. Critical values were calculated as the 95th percentile of the empirical distribution of the simulated wavelet coherencies following Grinsted et al. (2004) and Maraun and Kurths (2004).

Amongst the Western rivers (Figures 7(a), 7(b)) the coherency for the rivers Nith, Kelvin and Clyde has a similar structure, with statistically significant correlation around 1987-1990 and then again during 1998-2001 on the 1 year band, plus a period of high correlation between scales 4-6 years towards the end of the record in the River Kelvin. In terms of the phase difference, river flow and NAO seem to be out of phase, but this relationship is not constant over time. For the 1 year band, the phase difference changes from about the mid 1980s onwards, while for the 2 year band, it looks as if the phase difference can be divided into 3 different stages. The River Ewe, on the other hand, shows a correlation structure quite different from the rest of the Western rivers, with statistically significant correlation around 1987-1990 and around 1998-2001 (although the latter now extends up to 2004) in the 1 year band plus statistically significant correlation at higher scales (2-4 years band). The phase difference for this river is similar to the previous ones on the 1 year band, but considerably different at higher scales.

Amongst the rivers in the East (Figures 7 (c), 7(d)), the rivers Tweed, Water of Leith and Tay also show high correlation on the 1 year band around 1987-1990 and 1998-2001, plus a period of significant correlation on the 4 year band that can also be seen for the River Tay. The River Lossie seems to have a different pattern, with patches of high correlation along the 1 year band around 1991-1994 and 2002, ie, with a delay of about 1 year with respect to the rest of the rivers. All five Eastern rivers show intermittent periods of high correlation at higher scales. As was the case for the rivers in the West, NAO and river flow maxima are out of phase but the phase difference is not constant for every scale and time point.

To gain a better understanding of the phase difference, the 1 year band filtered NAO series was added to Figure 4. It is clear, for example from the plot for the River Nith, that the NAO series slowly shifted from being completely out of phase around 1976-1977 to being completely in phase at the beginning of the 1990s. A similar plot was produced for the AMO series, but the variability for the 1 year band filtered series was too small compared to the variability of the river maxima series to be informative and hence the plot is not shown here.

The correlation structure between river flow and AMO (Figure 8) resembles that of the NAO, although the relationship seems to be stronger than for the NAO. As before, there is moderate correlation for most of the time and scales, but it is not constant. The highest correlation is concentrated on the 1 year band, suggesting that AMO has a strong influence on the seasonal cycle of river maxima. A peak of high, significant correlation can be found for most of the rivers around 1987-1990, independently of whether they are in the East or West of the country. It is interesting that the AMO influence appears to be stronger in the second half of the record (from 1987 onwards) than it is in the first. Among the Western rivers (Figures 8(a), 8(b)), the Nith and Clyde present a patch of high correlation between 2000 and 2005, a feature that is also present on the Eastern rivers Dee, Tay and Tweed. The catchment area of

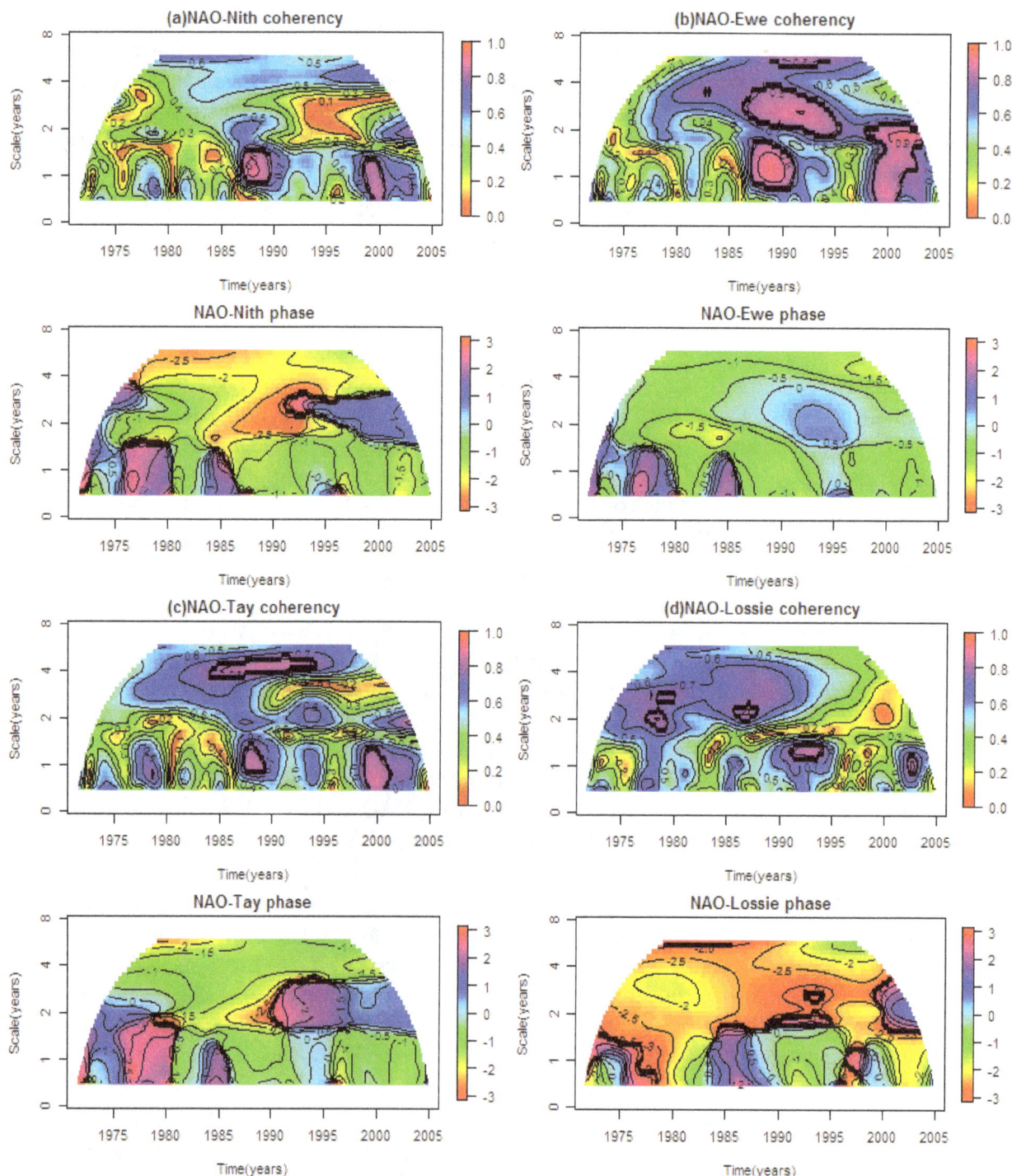

Figure 7: Wavelet coherency (top) and phase (bottom) between NAO and rivers (a)Nith, (b)Ewe, (c)Tay and (d)Lossie. The thick black contour lines on the wavelet coherency plots denote regions of statistically significant correlation.

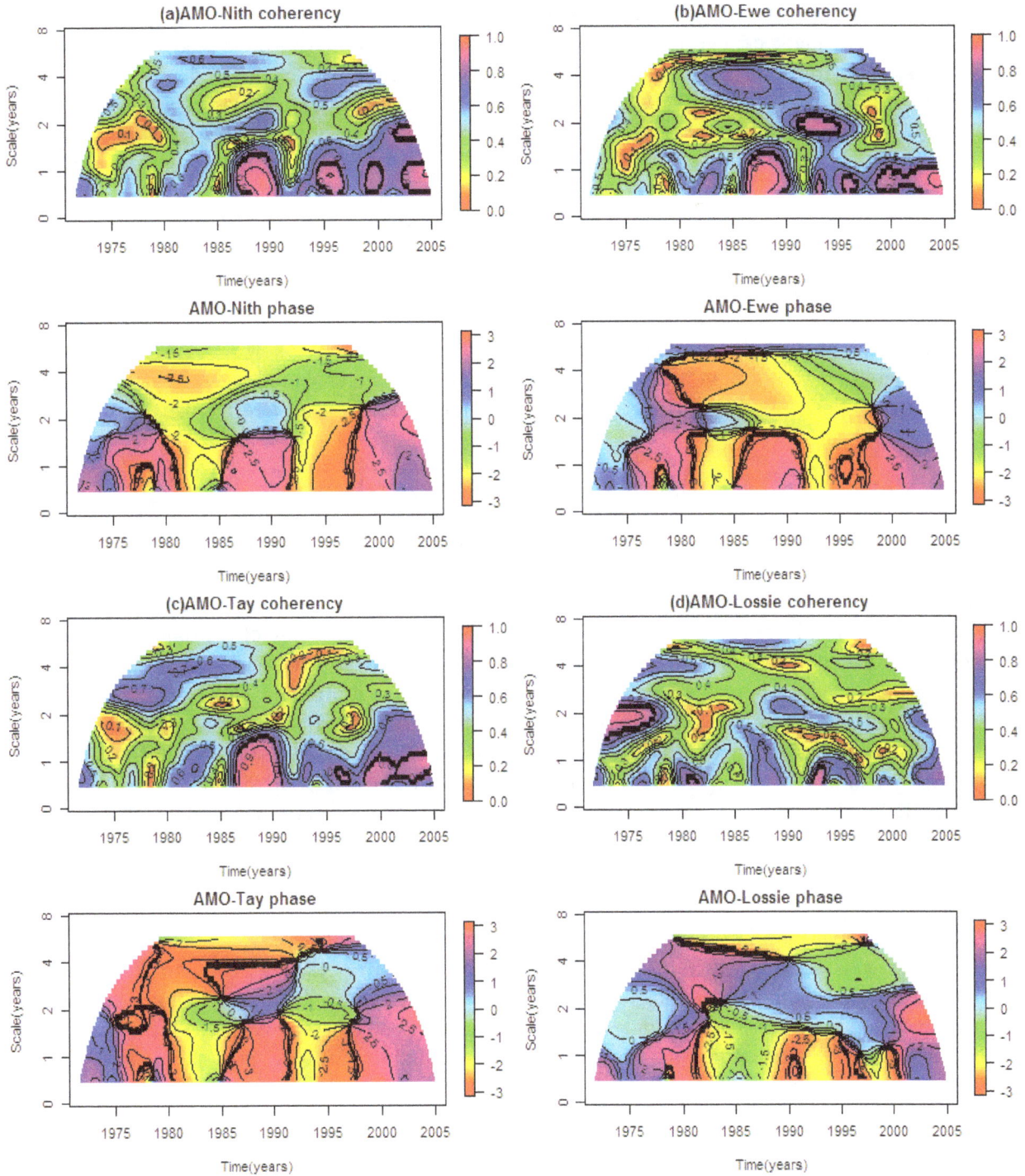

Figure 8: Wavelet coherency (top) and phase (bottom) between AMO and rivers (a)Nith, (b)Ewe, (c)Tay and (d)Lossie. The thick black contour lines on the wavelet coherency plots denote regions of statistically significant correlation.

these five rivers is much larger than that of the remaining four rivers, suggesting that the AMO influence might depend on the size of the catchment. Two rivers show slightly different patterns; these are the River Lossie, for which the correlation in the 1 year band appears to be higher (and significant) around 1993 and at the end of the record and at the beginning on the 2 year band, and the Water of Leith, which still shows high correlation in 1987-1990 but does not appear to be influenced by AMO as much as the rest of the rivers at the end of the record.

Peaks of significant correlation following Maraun and Kurths (2004) and Grinsted *et al.* (2004) have been identified at localized time periods and scales. While the correlation structures (as in coherency plots) between river maxima and AMO and NAO have common features (meaning that they are mainly concentrated on the 1 year band and that they are not constant over time) the phase difference plots are distinct. Even though river flow and AMO are also out of phase and the phase difference is not constant, such difference along the 1 year band appears to be nearly the opposite of what was observed for the NAO (Kerr (2000)).

4. Discussion and Future Work

Wavelet analysis is presented here as a powerful tool for exploring the variability and cross-correlation of non-stationary time series as well as comparing rivers at different locations. The results suggest differences between the East and the West in terms of long term trends and seasonal variability, the latter being higher for rivers in the West of Scotland. A clear indication of non-stationarity was found for for both eastern and western rivers. 1986 was detected as a "change point", when the seasonal variability is minimum. This is in agreement with Black (1996), who suggest a shift towards a "flood rich" period in the late 1980s. The 'cluster' of high variability just before that (from about 1977 to 1986), especially in the West, would correspond to the wettest period on record for the UK (Marsh (1995)). Figure 5(c) also suggest a North to South difference, as the rivers Water of Leith and Tweed, situated in the Southeast of Scotland, seem to have higher variability than those in the Northeast (Lossie, Dee and Tay).

Relationships between NAO and river flow have been reported in previous studies, either on a worldwide scale or restricted to Europe, while literature relating AMO to river flow or rainfall is less abundant. The relationship between NAO and river flow is not a simple one. In Europe, results from a cross-wavelet analysis suggested strong correlation with river flow during 1900-1950 on a scale band of 8-15 years (Labat (2010)). Shorthouse and Arnell (1997) investigated such a relationship during the period 1961-1990 looking at 744 river basins across Europe, finding spatial and temporal patterns in the NAO influence. However, their results, despite being indicative of what the relationship might be, must be interpreted carefully, for the conclusions are based on the results of a Pearson correlation analysis, which assumes independence of observations (something that is rarely the case when dealing with time series) and a linear relationship between the two series. Wavelet based cross-correlation appears to be a good alternative to traditional correlation, even though some authors (Torrence and Compo (1998)) argue that smoothing is somehow contrary to the purpose of using wavelets, which is that of improving the localization of events, as localization is decreased by smoothing. Nonetheless, cross-correlation is highly informative and allows exploration of relationships in a broader time frame, with the assumption of independence no longer being made. With respect to the linearity of the relationship, the interpretation is less clear. While some authors define coherency as the strength of the linear relationship between two series, it has been suggested that the phase difference might be interpreted not only as the time lag but also the degree of linearity (Velasco and Mendoza (2008)); a value close to the extremes $-\pi$ and π would suggest a (negative/positive)linear relationship, while values in between would mean that the relationship is not linear.

In an effort to investigate how current climate change might impact on the frequency of extreme events, relationships between NAO and AMO and river maxima were explored. The results from the wavelet coherency analysis suggest that the influence of NAO varies slightly from catchment to catchment, with Eastern rivers showing periods of correlation at scales higher than the 1 year band, the latter being common to most of the rivers in both the East and West. The relationship between the AMO and river maxima also changes slightly from catchment to catchment, but the influence appears to be stronger than that of the NAO, especially at the end of the 1980s. During this period of high correlation, the AMO has been said to be in negative or cold phase, although unusually high values (to be in a cold phase) were recorded, regarded as a period of 'transition' between the cold and warm phase. From that point onwards, the correlation with river maxima is higher, suggesting that rivers might be more affected when AMO is in its warm phase. At this point, this results should be interpreted as merely indicative, and further research is needed to draw definite conclusions. The AMO has an oscillation period of about 60 years and here we are looking at possible relationships over 30 years. It would be useful to extend the length of the records to see how this oscillation has influenced river flow in the long term, to investigate whether the effect has always been similar or has changed through time. Further, AMO is claimed to be linked to summer rather than winter climate (Knight et al. (2006); Sutton and Hodson (2005)). A seasonal analysis to gain a better understanding of the relationship would be informative.

The NAO is now "... well recognized to have an important influence on European climate and its variability" (Marković and Koch (2005)) particulary in the winter months (Marković and Koch (2005); Shorthouse and Arnell (1997)). Since the NAO has such a strong influence on rainfall, the expectation is that this influence will extend to river flow too. Previous studies have linked it to winter river discharge globally (Dettinger and Diaz (2000); Labat (2010)) and in Europe (Shorthouse and Arnell (1997); Kingston et al. (2009); Bouwer et al. (2008)). Macklin and Rumsby (2007) investigated the relationship between NAO and floods in upland catchments in the UK over the last 250 years, arguing that "... the non-stationarity of the flood series was related to decadal and multi decadal scale climatic fluctuations" (Macklin and Rumsby (2007)). They claim that the relationship changes both spatially and temporally, with the NAO influence over Scotland being different to that in England and Wales (Macklin and Rumsby (2007)) and within Scotland itself, where 3 upland regions were studied, Glencoe, An Teallach, in the West of Scotland and the Cairngorms, in the East. Their results suggest an increase in the frequency of floods in the second part of the 19th century and again in the 1980s, the latter associated with high rainfall (autumn) and positive NAO (Macklin and Rumsby (2007)) (Figure 6(a)), contrary to data from England, which suggest higher frequency floods when NAO is negative (Macklin and Rumsby (2007)). In agreement with the findings of Macklin and Rumsby (2007), Shorthouse and Arnell (1997) report strong positive correlation between regional runoff and NAO in Scotland, especially in winter (December,January,February), although the strength of the correlation varies across the country.

Even though the AMO has only been defined recently, its oscillatory nature was observed in the early 1970s (Kerr (2000)), when researchers noticed an increase in the North Atlantic sea surface temperature during 1910-1940 that was accompanied with an increase in global air temperature, after which a phase of cooling down began, both in sea surface and global air temperature, to then warm up again in the 1980s. Phases of warm/cold AMO index have been related to anomalous regional climate, particularly in the North West of Europe. This dependence seems to vary seasonally, with the highest impact during the summer months (June, July and August) (Knight *et al.* (2006); Sutton and Hodson (2005)).

The rivers Lossie and Ewe seem to be affected differently from the rest of the rivers investigated in this study; they are both in the North of Scotland and have relatively small catchments. This might be an indication of a North to South differentiation that had already been suggested by Kingston *et al.* (2009) and possibly a catchment size effect, reinforcing the importance of carrying studies on a regional level (Government (2010)).

Differences between the East and the West of Scotland have been identified in river flow maxima over the last 40 years, both in the long term trend and seasonal pattern. The latter appears to be subject to great variability, being slightly higher in the West than in the East, with a common time point around 1986 for all the rivers, when the variability is minimum. Periods of greater seasonal variability (and hence a more pronounced seasonal pattern) coincide with flood rich periods. Some of these changes might be explained through the influence of external climatic drivers such as the AMO and the NAO. The correlation strength between river maxima and these two indices varies temporally, appearing to be stronger, specially for the AMO, in the second half of the record (from about 1987 onwards). The influence of AMO and NAO not only changes through time but also the phase difference is highly variable, and is different depending on location (East/West) but also on the size of the catchment. This analysis could be extended wider to cover Europe, in order to gain a better understanding of the spatial dependence in extreme river flows, which will contribute to the improvement of flood risk management.

5. Acknowledgements

We would like to thank the Scottish Environment Protection Agency (SEPA) and the National River Flow Archive (NRFA) for providing data, Alistair Cargill (SEPA) for discussion, and the University of Glasgow Kelvin-Smith studentship scheme for funding. We also thank an anonymous referee who helped improve the manuscript.

References

Black A (1996). "Major Flooding and Increased Flood Frequency in Scotland since 1988." *Physics and Chemistry of the Earth*, **20**(5-6), 463–468.

Black A, Burns C (2002). "Re-assessing the Flood Risk in Scotland." *The Science of the Total Environment*, **294**, 169–184.

Bouwer L, Vermaat J, Aerts J (2008). "Regional Sensitivies of Mean and Peak River Discharge to Climate Varability in Europe." *Journal of Geophysical Rsearch*, **113**, D19103.

Daubechies I (1992). *Ten Lectures on Wavelets*. Philadelphia: SIAM.

Delworth T, Mann M (2000). "Observed and Simulated Multidecadal Variability in the Northern Hemisphere." *Climate Dynamics*, **16**, 661–676.

Dettinger M, Diaz H (2000). "Global Characteristics of Stream Flow Seasonlity and Variability." *Journal of Hydrometeorology*, **1**, 289–310.

Government S (2010). Electronic Resource [Accessed 13/09/2010]. URL http://www.scotland.gov.uk/Topics/Environment/Water/Flooding/FRMAct

Gregory J, PD J, Wigley T (1991). "Precipitation in Britain: an Analysis of Area-average Data Updated to 1989." *International Journal of Climatology*, **11**, 331–345.

Grinsted A, Moore J, Jevrejeva S (2004). "Application of the Cross Wavelet Transform and Wavelet Coherence to Geophysical Time Series." *Nonlinear Processes in Geophysics*, **11**, 561–566.

Hurrell J (1995). "Decadal Trends in the North Atlantic Oscillation: Regional Temperatures and Precipitation." *Science*, **269**, 676–679.

Jenkins GJ, Murphy JM, Sexton DMH, Lowe JA, Jones P, Kilsby CG (2009). "UK Climate Projections: Briefing Report." *Technical report*, Met Office Hadley Centre, Exeter, UK.

Kerr R (2000). "A North Atlantic Climate Pacemaker for the Centuries." *Science*, **288**(5473), 1984–1985.

Kingston D, Hannah D, Lawler D, McGregor G (2009). "Climate-river Flow Relationships across Montane and Lowland Environments in Northern Europe." *Hydrological Processes*, **23**, 985–996.

Knight J, Folland C, Scaife A (2006). "Climate Impacts of the Atlantic Multidecadal Oscillation." *Geophysical Reserach Letters*, **33**, L17706.

Labat D (2005). "Recent Advances in Wavelet Analyses: Part 1. A Review of Concepts." *Journal of Hydrology*, **314**, 275–288.

Labat D (2010). "Cross Wavelet Analyses of Annual Continental Frewshwater Discharge and Selected Climate Indices." *Journal of Hydrology*, **385**, 269–278.

Labat D, Ronchail J, Guyot J (2005). "Recent advances in wavelet analyses: Part 2 - Amazon, Parana, Orinoco and Congo discharges time scale variability." *Journal of Hydrology*, **314**, 289–311.

Macklin M, Rumsby B (2007). "Changing Climate and Extreme Floods in the British Uplands." *Transactions of the Institute of British Geographers*, **32**, 168–187.

Maraun D, Kurths J (2004). "Cross Wavelet Analysis: Significance Testing and Pitfalls." *Nonlinear Processes in Geophysics*, **11**, 505–514.

Marković D, Koch M (2005). "Wavelet and Scaling Analysis of Monthly Precipitation Extremes in Germany in the 20_{th} Century: Interannual to Interdecadal Oscillations and the North Atlantic Oscillation Influence." *Water Resources Research*, **41**, W09420.

Marsh T (1995). "The 1995 Drought - a Water Resources Review in the Context of the Recent Hydrological Instability." *Hydrological Data UK Series*, pp. 25–33.

Mayes J (1996). "Spatial and Temporal Fluctuations of Monthly Rainfall in the British Isles and Variations in the Mid-latitude Westerly Circulation." *International Journal of Climatology*, **16**, 585–596.

Percival D, Mofjeld H (1997). "Analysis of subtidal coastal sea level fluctuations using wavelets." *Journal of the American Statistical Association*, **92**(439), 868–880.

Percival D, Walden A (2006). *Wavelet Methods for Time Series Analysis*. Cambridge Series in Statistical and Probabilistic Mathematics. Cambridge University Press.

Rossi A, Massei N, Laignel B, Sebag D, Copard Y (2009). "The response of the Mississipi River to climate fluctuations and reservoir construction as indicated by wavelet analysis of streamflow and suspended-sediment load, 1950-1975." *Journal of Hydrology*, **377**, 237–244.

Sen A (2009). "Spectral-temporal characterization of riverflow variability in England and Walse for the period 1865-2002." *Hydrological Processes*, **23**, 1147–1157.

Shorthouse C, Arnell N (1997). "Spatial and Temporal Variability in European River Flows and the North Atlantic Oscillation." *FRIEND'97 - Regional Hydrology: Concepts and Models for Sustainable Water Resource Management*, **246**, 77–85.

Smith L, Turcotte D, Isacks B (1998). "Stream flow characterization and feature detection using a discrete wavelet transform." *Hydrological Processes*, **12**, 233–249.

Sutton R, Hodson D (2005). "Atlantic Ocean Forcing of North American and Europe Summer Climate." *Science*, **309**, 115–118.

Torrence C, Compo G (1998). "A Practical Guide to Wavelet Analysis." *Bulletin of the American Meteorological Society*, **79**, 61–78.

Torrence C, Webster P (1999). "Interdecadal changes in the ENSO-Monsoon system." *Journal of Climate*, **12**, 2679–2690.

Velasco V, Mendoza B (2008). "Assessing the Relationship between Solar Activity and Some Large Scale Climatic Phenomena." *Advances in Space Research*, **42**, 866–878.

Werritty A (2002). "Living with Uncertainty: Climate Change, River Flows and Water Resource Management in Scotland." *The Science of the Total Environment*, **294**, 29–40.

White W (2001). "Water in rivers: flooding." *Proceedings of the Institution of Civil Enginners - Water, Maritime and Energy*, **148**(2), 107–118.

Young J, Davies (1989). "The Realistic Criteria for Flood-control Design." *Hydrology and water resources symposium 1989: comparisons in austral hydrology, institution of engineers, australia, national conference publications*, **89**, 227–231.

Affiliation:

Maria Franco-Villoria
School of Mathematics and Statistics, University of Glasgow
15 University Gardens, G12 8QW, Glasgow
E-mail: `m.franco-villoria.1@research.gla.ac.uk`

10

Use of the Dagum Distribution for Modeling Tropospheric Ozone levels

Benjamin Sexto M.
Colegio de Postgraduados

Humberto Vaquera H.
Colegio de Postgraduados

Barry C. Arnold
UC, Riverside

Abstract

This paper deals with the use of the Dagum distribution to model the maximum daily levels of tropospheric ozone. We compare the fit of the Dagum distribution against the Generalized Extreme Value distribution (GEV) by using the Kolmogorov-Smirnov test and the Akaike criterion for model selection. Also we propose a methodology for estimating long term trends in the daily maxima of tropospheric ozone by using the Vector Generalized Linear Model (VGLM) and quantiles of the Dagum distribution. Ozone data from Pedregal Station in Mexico City (one with the worst air pollution in the World) are analyzed for the period 2001-2008. Results show that the Dagum model has a similar or better fit than the GEV model. The quantiles of Dagum distribution and VGLM show evidence of a downward trend in high ozone levels at Pedregal Station.

Keywords: trends, urban ozone, extreme value, vgam.

1. Introduction

One important pollutant in big cities is ozone (O_3) which in high levels (above .12 ppm) is harmful to human health (Ebi and McGregor 2008). Urban ozone effects may be more severe in certain susceptible groups such as children, elderly, sick people and people who enjoy outdoor exercise (Ponce de Leon, Anderson, Bland, and Bower 1996).

In tropospheric ozone data analysis the traditional distributions used are the Generalized Extreme Value distribution (GEV) and the Pareto distribution.

Extreme values in environmental time series are important because of their applicability to the analysis of catastrophic phenomena such as extreme ozone observations, and extreme meteorological conditions (floods, winds, temperature, etc). The statistics of extremes can undoubtedly be useful in applications relating to distributions with light or bounded tails, but they are found to be most useful for variables that have a heavy tailed distribution (Katz,

Parlange, and Naveau 2002).

The Dagum distribution has two parameters, one of shape and the other of scale. This distribution has been used by economists as a distribution for modeling country incomes because of its property of having a heavy right tail. Mielke (1973) used the Kappa distribution (with three parameters) to model the amount of rainfall precipitation. The Kappa distribution Mielke and Johnson (1974) includes the Dagum distribution in a different parametrization (referred as the Beta-K distribution). Dagum (1977) and Fattorini and Lemmi (1979) proposed the Kappa distribution as an income distribution.

In this paper use of the Dagum distribution is proposed for modeling daily maximum levels of ozone at a specific location. Subsequently, the fit of the Dagum distribution and that of the generalized extreme value distribution (GEV) are compared. An additional goal is to propose methodology for estimating long term trends in the daily maxima of tropospheric ozone, using information from the environmental monitoring station in Pedregal, Mexico City.

1.1. Dagum distribution

The Dagum distribution is a heavy-tailed distribution developed by Camilo Dagum in the 70's for modeling income distributions as an alternative to the Pareto (Pareto 1895) and log-normal (Gibrat 1931) models. The most general form of the Dagum distribution has the following cumulative distribution function.:

$$F(x) = \alpha + (1 - \alpha)[1 + (x/b)^{-a}]^{-p} \tag{1}$$

The Dagum distributions of Type I, II and III correspond to cases where $\alpha = 0$, $0 < \alpha < 1$ and $\alpha < 0$ respectively. The Dagum type II distribution was proposed as a model for income distribution allowing for zero or negative income. It seems especially appropriate for wealth data, where there are often a large number of economic units with zero net assets. The Dagum distribution of Type III is associated with a positive lower limit for X, x_0. In this paper we will work with the Dagum of type I. Henceforth this distribution will be simply referred as the Dagum distribution. The Dagum distribution is a special case of the generalized beta distribution of the second kind (GB2). The density of the GB2 distribution is:

$$f(x) = \frac{ax^{ap-1}}{b^{ap}B(p, q)\left[1 + (x/b)^a\right]^{p+q}}, \quad x > 0 \tag{2}$$

where $b > 0$ is the scale parameter and $a > 0$, $p > 0$, $q > 0$ are the shape parameters. In (2), if the shape parameter q is set equal to 1, the Dagum density is obtained:

$$f(x) = \frac{apx^{ap-1}}{b^{ap}\left[1 + \left(\frac{x}{b}\right)^a\right]^{p+1}}, \quad x > 0 \tag{3}$$

where $a, b, p > 0$. The Dagum distribution function has a closed form:

$$F(x) = \left[1 + \left(\frac{x}{b}\right)^{-a}\right]^{-p}, \quad x > 0 \tag{4}$$

where $a, b, p > 0$. The parameter b is the scale parameter, while a and p are shape parameters.

In the case in which $ap > 1$ the density has an interior mode. The mode for Dagum distribution is:

$$x_{mode} = b \left(\frac{ap - 1}{a + 1} \right)^{1/a} \tag{5}$$

The quantile function also has a closed form:

$$F^{-1}(u) = b \left[u^{-1/p} - 1 \right]^{-1/a}, \quad for \ 0 < u < 1 \tag{6}$$

The k-th moment exists for $-ap < k < a$ as follows:

$$E(X^k) = \frac{b^k \Gamma \left(p + k/a \right) \Gamma \left(1 - k/a \right)}{\Gamma(p)} \tag{7}$$

where $\Gamma(\cdot)$ denotes the Gamma function.

In particular the mean and variance are:

$$E(X) = \frac{b \Gamma \left(p + 1/a \right) \Gamma \left(1 - 1/a \right)}{\Gamma(p)} \tag{8}$$

$$var(X) = \frac{b^2 \left[\Gamma(p) \Gamma \left(p + 2/a \right) \Gamma \left(1 - 2/a \right) - \Gamma^2 \left(p + 1/a \right) \Gamma^2 \left(1 - 1/a \right) \right]}{\Gamma^2(p)} \tag{9}$$

In practical situations the estimated value of parameter a is usually small (in economic applications a gets smaller as income inequality increases) (Dagum and Lemmi 1989).

Parameter estimation can be implemented using the method of maximum likelihood. Let $X_1, ..., X_n$ be a random sample of size n from the Dagum distribution, the log-likelihood function is defined as:

$$\ell = n \log a + n \log p + (ap - 1) \sum_{i=1}^{n} \log x_i - nap \log b - (p + 1) \sum_{i=1}^{n} \log \left[1 + \left(\frac{x_i}{b} \right)^a \right] \tag{10}$$

A variety of standard optimization programs can be used to maximize this function. In particular, the package EVIR in R can be utilized.

1.2.   GEV distribution

Three types of extreme value limit distributions play a fundamental role in the analysis of extremes of environmental data: Fréchet, Weibull and Gumbel. The Generalized Extreme Value (GEV) is a combination of these three types of extreme value limit distributions (von Mises (1936, 1954) and Jenkinson (1955)), and its distribution function is:

$$G(x; \mu, \sigma, \xi) = \exp \left\{ - \left[1 + \xi \left(\frac{x - \mu}{\sigma} \right) \right]_+^{-\frac{1}{\xi}} \right\} \tag{11}$$

where $\sigma > 0$, $-\infty < \mu < \infty$, $1 + \xi(x - \mu)/\sigma > 0$, $x_+ = max\{x, 0\}$. The parameter μ is a location parameter, σ is a scale parameter, and ξ shape parameter. The cases in which $\xi > 0$, $\xi < 0$ and $\xi = 0$, correspond to the Fréchet, Weibull and Gumbel distributions respectively. The quantile function of the GEV distribution is:

$$G^{-1}(u) = \mu - \frac{\sigma}{\xi} \left[1 - \{ -\log u \}^{-\xi} \right] \tag{12}$$

with $0 < u < 1$. The value $G^{-1}(1 - u)$ is the return level associated with the return period $1/u$.

2. Statistical Methodology

2.1. Ozone Data

The pollutant concentrations to be studied correspond to an urban site located South of Mexico City. These measurements are integrated in the Air Quality Monitoring Network of the Valle de Mexico Metropolitan Area, managed by the Atmospheric Monitoring System (SIMAT) of the Mexico City Government. The analyzed data correspond to daily maxima of ozone measures (ppm). Ozone concentrations were monitored using UV absorption photometry using the API 400 and API 400A. The study data correspond to the period from 2001 to 2008. The data is available at http://www.sma.df.gob.mx/simat2/informaciontecnica.

2.2. Block Maxima

For the ozone data set, a block length of three days is considered to be a long enough period of separation between observations to achieve independence. In each block, the maxima was obtained (Block maxima). The Block maxima method is described by (Gaines and Denny 1993). The distribution of maximum ozone levels is not the same each year; there is a trend towards lower peak levels of ozone over the years, thus the series is not stationary. Therefore, it is appropriate to analyze and adjust the maximum levels of ozone for each year to minimize the non-stationarity problem. A time series plot for block maxima of ozone levels is in Fig.(1). Autocorrelation in ozone data would have little effect on the bias of parameter estimates, but their variance is affected Vaquera H (1997). The use of block maxima reduces the undesirable consequences of the autocorrelation.

2.3. Parameter estimation for GEV and Dagum

In the case of Dagum distribution, the Parameter values for $(\hat{a}, \hat{b}, \hat{p})$ that maximize the log-likelihood were obtained using a computational routine in the "VGAM" library for R. The calculation of estimates of the parameters of the GEV model was implemented using the maximum likelihood method with the *EVIR* package for R.

Smith (1985) observed that, for the GEV distribution (11) in the case in which $\xi < -1/2$, the usual asymptotic distributional properties of maximum likelihood estimators (MLE) do not hold. In contrast, in the case of the Dagum distribution the maximum likelihood estimators do not have such problems according to Kleiber and Kotz (2003). Consequently, if the two models, GEV and Dagum, provide comparable fits to a given data set, an argument can be advanced in favor of using the Dagum model. As we shall see, this is the case for the Ozone data analyzed here.

2.4. Assessing the of Fit of the Ozone data

The Kolmogorov-Smirnov statistic was used for comparing the Dagum and GEV distribution fits of the ozone maxima time series for each year in the range 2001-2008. The test statistic

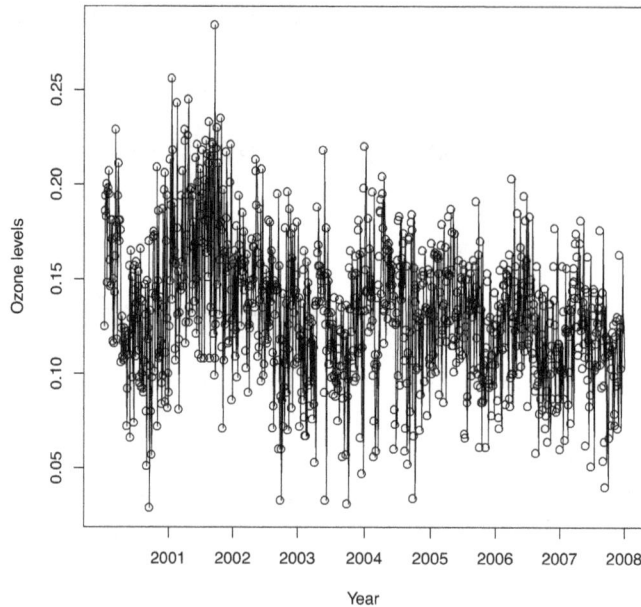

Figure 1: Daily maxima ozone levels for Pedregal (ppm)

is:

$$D = \sup_{-\infty < x < \infty} |F_n(x) - F_0(x)|$$

where $F_n(\cdot)$ is the empirical distribution function and $F_0(\cdot)$ is the fitted distribution function with parameters estimated by maximum likelihood.

Another criteria for assessing the fit is the Akaike Information Citerion (AIC) Akaike (1974). When comparing models using the maximum likelihood method for fitting, the AIC is calculated using the following expression:

$$AIC(k) = 2k - 2log(L(\hat{\theta})) \tag{13}$$

where, k is the number of model parameters estimated by the method of maximum likelihood and $L(\hat{\theta})$ is the likelihood function evaluated at the maximum likelihood estimate $\hat{\theta}$. The preferred model will be the one with the lowest AIC.

2.5. Trend estimation in the Ozone data levels

Quantile estimates

Reyes, Vaquera, and Villaseñor (2010) proposes a statistical methodology to analyze the trends of very high values of tropospheric ozone, the methodology is based on the estimation of percentiles of the distribution of extreme values (GEV). In this work a similar idea is used for investigating trends for Dagum and GEV distribution. For the calculation of the quantile estimates, we can use maximum likelihood method with the **EVIR** package in R.

Vector Generalized Linear Model

The Vector Generalized Linear Model (VGLM) allows us to determine if there is a linear relationship between the parameters of the Dagum distribution with time as a covariate. The estimation of the parameters of the VGLM model can be performed in the VGAM library of R using the $VGLM()$ function. The VGLM and VGAM were introduced by Yee and Hastie (2003) and Yee and Wild (1996).

The VGAM/VGLM are implemented in the package VGAM (Yee 2007), working in R. The VGAM and VGLM allow all parameters of the distribution be modeled as linear or smoothed functions of covariates. Suppose the observed response y is a q-dimensional vector. The VGLM is defined as a model for which the conditional distribution of Y given the explanatory variables x is of the form:

$$f(y|x; B) = h(y, \eta_1, ..., \eta_M) \tag{14}$$

for some known function $h(\cdot)$, where $B = (\beta_1, \beta_2, ..., \beta_M)$ is a $p \times M$ matrix of unknown regression coefficients, and the j-th liner predictor is:

$$\eta_j = \eta_j(x) = \beta_j^T x = \beta_{(j)1}x_1 + \cdots + \beta_{(j)p}x_p = \sum_{k=1}^{p} \beta_{(j)k}x_k, \quad j = 1, ..., M \tag{15}$$

where $x = (x_1, ...x_p)^T$ with $x_1 = 1$ if there is an intercept. In our case the covariate is time.

The VGAM provide extensions to VGLM additive models, the equation predictor is generalized to a sum of smoothed functions of the individual covariates:

$$\eta_j(x) = \beta_{(j)1} + f_{(j)2}(x_2) + \cdots + f_{(j)p}(x_p) = \beta_{(j)1} + \sum_{k=2}^{p} f_{(j)k}(x_k), \quad j = 1, ..., M \tag{16}$$

The η_j are referred to as additive predictors. $f_k = (f_{(1)k}(x_k), ..., f_{(M)k}(x_k))$ is focused on uniqueness, and are estimated simultaneously using "vector smoothers". VGLM are usually estimated by maximum likelihood using Fisher scoring or Newton-Raphson.

3. Results for Mexico City Ozone levels

The visual comparison of the adjustment of the Dagum distribution with the empirical distribution and the corresponding adjustment of the GEV distribution for maximum daily ozone levels per year is found in Figure 2. From table 1 we can observe a very similar fit for the Dagum and GEV models, and in particular in the years 2003, 2004, 2005 and 2008 the Dagum fit appears to be somewhat better.

In table 1, the p-values for the Kolmogorov-Smirnov test and the Akaike values (AIC) are shown. We can conclude that maximum daily ozone observations are satisfactorily modeled by both the Dagum distribution and the Generalized Extreme Value distribution. Table 1 shows that Dagum distribution fits better in 2003, 2005 and 2008 according to both the Kolmogorov and Akaike criteria.The Overall AIC in all years(2001-2008) for Dagum was -3616.972, and for GEV -3613.

In figure(1), a non stationary pattern in the Ozone series is clear. For this reason, an analysis has been implemented for each year separately. In table 2, the analyses for the years indicates that there is a perceptible trend in the behavior of the estimated parameters a and b of the Dagum model.

2008

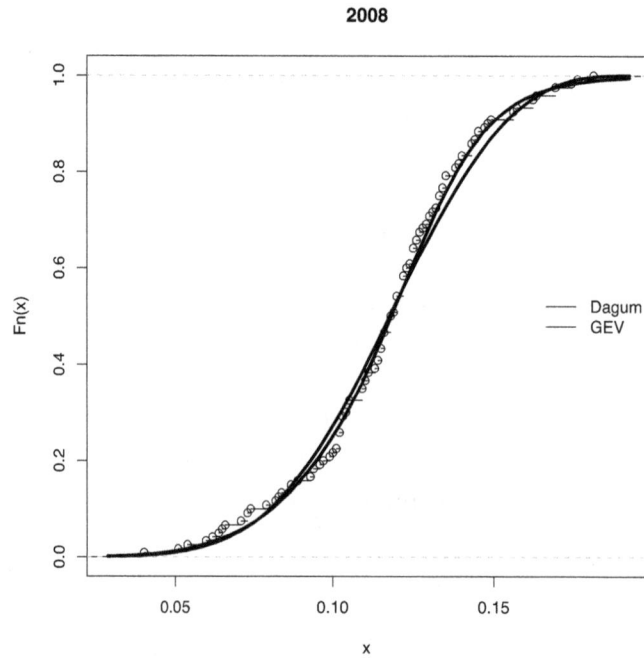

Figure 2: Example Fit Dagum vs GEV:2008

Table 1: Kolmogorov $p - values$ and AIC Akaike statistic

Year	$p - value$		AIC	
	Dagum	GEV	Dagum	GEV
2001	0.6371	0.7933	-426.6164	-434.7164
2002	0.5414	0.7623	-409.4166	-416.752
2003	0.9728	0.801	-475.4264	-474.6734
2004	0.9545	0.9743	-478.3866	-476.224
2005	0.6972	0.2813	-460.532	-456.9844
2006	0.5107	0.6088	-501.6258	-508.6648
2007	0.8596	0.9089	-502.811	-505.7758
2008	0.9318	0.5637	-518.5954	-516.0032

Table 2: Estimated parameters of Dagum Distribution

Year	Parameter		
	a	b	p
2001	8.3479	0.1615	0.4202
2002	9.7786	0.1959	0.4150
2003	13.5001	0.1669	0.2918
2004	11.2898	0.1486	0.2999
2005	14.1866	0.1701	0.2371
2006	14.9753	0.1563	0.2679
2007	10.3362	0.1394	0.4608
2008	12.3889	0.1369	0.3520

The observed change in the parameters over the years is in accordance with the results obtained using a vector generalized linear model (VGLM) in which year is a covariable as follows:

$$\log(a) \quad = \quad \eta_1 = \beta_{(1)1}x_1 + \beta_{(1)2}x_2$$
$$\log(b) \quad = \quad \eta_2 = \beta_{(2)1}x_1 + \beta_{(2)2}x_2$$
$$\log(p) \quad = \quad \eta_3 = \beta_{(3)1}x_1 + \beta_{(3)2}x_2$$

where $x_1 = 1$ corresponding to the intercept, x_2 is the time (years), $q = 1$ and $M = 3$. The results are presented in table 3.

Table 3: Dagum regression coefficients

Coefficients	Value	Std. Error	t-value
(Intercept):1	2.087559	0.1351299	15.44854
(Intercept):2	-1.676371	0.0367075	-45.66831
(Intercept):3	-0.920553	0.2083267	-4.41879
year:a	0.06091	0.0271134	2.24648
year:b	-0.036772	0.0065516	-5.61265
year:p	-0.022878	0.041165	-0.55575

Considering a significance level $\alpha = 0.05$ from table 3, we observe a significant linear trend in the estimated coefficients a and b (year:1 and year:2). Also the signs of these regression coefficients are consistent with the estimates from table 2. In the case of parameter p there is no significant linear trend.

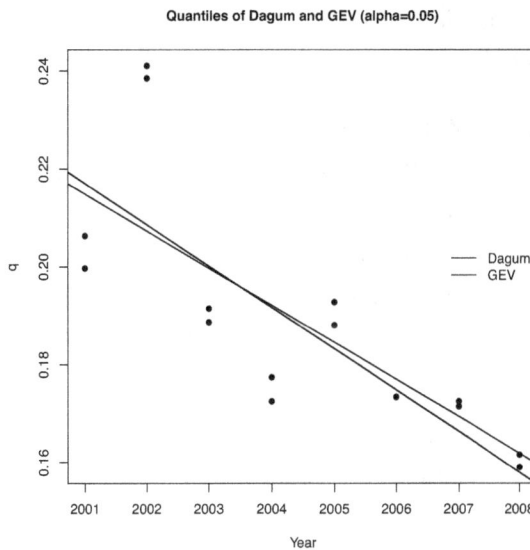

Figure 3: $(1 - \alpha)100$ quantiles of Dagum and GEV

Table 4 shows the $(1 - \alpha)100$ quantiles of Dagum and GEV distributions for each of the years with α=0.05 and 0.10. A downward linear trend is thus observed in these quantiles for high ozone levels.

A graphic representation is given in figure 3.

To test the proposed models for investigating trends in ozone levels, Dagum and GEV models were fitted for the period 2001-2006 and were used to forecast quantiles for the following two

Table 4: quantiles $(1 - \alpha)100$ of Dagum

Year	quantiles of Dagum		quantiles of GEV	
	0.05	0.10	0.05	0.10
2001	0.2062722	0.1877332	0.1996686	0.1865772
2002	0.2410239	0.2223855	0.2384297	0.2226471
2003	0.188601	0.177531	0.1914159	0.181634
2004	0.1724811	0.1604845	0.1774077	0.1651994
2005	0.1880305	0.1772189	0.192768	0.182186
2006	0.1734096	0.1641078	0.1733466	0.1648492
2007	0.1714448	0.1589807	0.1724634	0.1611932
2008	0.1590109	0.1490738	0.1615318	0.1529513

years. The forecasted 0.95 quantiles for 2007 and 2008 were (0.167, 0.161) respectively for the Dagum model and (0.190, 0.186) in same years for the GEV model. These forecasted quantiles are generally in agreement with the fitted quantiles in figure 3 (corresponding to the years 2001-2008).

In general, note that, as expected since both models fit the data well, the difference is small between the quantiles of both Dagum and GEV distributions, and they exhibit the same trend.

4. Conclusions

With regard to the implementation of the Dagum distribution to model extreme values in ozone levels, we can conclude that:

- Based on the Kolmogorov statistic and the Akaike criteria, the Dagum distribution provides similar and sometimes better fits than does the GEV distribution for the Pedregal ozone data.

- With the results obtained in this paper, we justify the implementation of the Dagum distribution to model extreme values. The Dagum distribution is an appealing option for modeling extreme events, when using maximum likelihood estimators since it does not have the distributional problems associated with the MLE's in the GEV model.

In addition, a downward trend with time was observed for maximum ozone levels and was confirmed by two relevant techniques:

- With the Vector Generalized Linear Model (VGLM), a trend was confirmed in the estimated parameters a and b of the Dagum distribution

- The estimated $(1 - \alpha)100$ quantiles corresponding to both of the Dagum and GEV models, exhibited a very similar downward trend as a function of time (years).

References

Akaike H (1974). "A new look at the statistical model indentification." *IEEE Transactions on Automatic Control*, **19**, 716–722.

Dagum C (1977). "A New Model for Personal Income Distribution: Specification and Estimation." *Economie Appliquée*, **30**, 413–437.

Dagum C, Lemmi A (1989). "A contribution to the analysis of income distribution and income inequality, and a case study: Italy." *Research on Economic Inequality*, **1**, 123–157.

Ebi KL, McGregor G (2008). "Climate Change, Tropospheric Ozone and Particulate Matter, and Health Impacts." *Environmental Health Perspectives*, **116-11**, 1449–1455.

Fattorini L, Lemmi A (1979). "Proposta di un modello alternativo per l'analisi della distribuzione personale del reddito." *Atti Giornate di Lavoro AIRO*, **28**, 89–117.

Gaines SD, Denny MW (1993). "The largest, smallest, highest, lowest, longest, and shortest: extremes in ecology." *Ecology*, **74**, 1677–1692.

Gibrat R (1931). *Les Inégalités Économiques*. Librairie du Recueil Sirey, Paris.

Jenkinson A (1955). "The frequency distribution of the annual maximum (or minimum) values of meteorological elements." *Quart. J. Roy. Meteo. Soc.*, **81**, 158–171.

Katz RW, Parlange MB, Naveau P (2002). "Statistics of extremes in hydrology." *Advances in Water Resources*, **25**, 1287–1304.

Kleiber C, Kotz S (2003). *Statistical Size Distributions in Economics and Actuarial Sciences*. John Wiley, Hoboken, New Jersey.

Mielke PW (1973). "Another family of distributions for describing and analyzing precipitation data." *Journal of Applied Meteorology*, **12**, 275–280.

Mielke PW, Johnson ES (1974). "Some generalized beta distributions of the second kind having desirable application features in hydrology and meteorology." *Water Resources Research*, **10**, 223–226.

Pareto V (1895). "La Legge della Domanda." *Giornale degli Economisti, English Translation in Rivista di Politica Economica*, **10, 87(1997)**, 59–68, 691–700.

Ponce de Leon A, Anderson H, Bland J, Bower J (1996). "Effects of air pollution on daily hospital admissions for respiratory disease in London between 1987-88 and 1991-92." *J Epidemiol Comm Health*, **Vol. 50 (Supplement 1)**, S63–S70.

R Development Core Team (2007). *R: A Language and Environment for Statistical Computing*. R Foundation for Statistical Computing, Vienna, Austria. ISBN 3-900051-07-0, URL http://www.R-project.org.

Reyes J, Vaquera H, Villaseñor J (2010). "Estimation of trends in high urban ozone levels using the quantiles of (GEV)." *Environmetrics*, **21**(5), 470–481.

Smith RL (1985). "Maximum likelihood estimation in a class of non-regular cases." *Biometrika*, **72**, 67–90.

Vaquera H H (1997). *On the statistical analysis of trend in tropospheric ozone levels*. Ph.D. thesis, Tulane University.

von Mises R (1936). "La distribution de la plus grande de n valeurs." *Revue Mathématique de l'Union Interbalkanique (Athens)*, **1**, 141–160.

von Mises R (1954). "La distribution de la plus grande de n valeurs." *American Mathematical Society*, **Selected Papers Volumen II**, 271–294.

Yee T (2007). *A Usert's Guide to the vgam Package*. URL `http://www.stat.auckland.ac.nz/~yee/VGAM`.

Yee T, Hastie T (2003). "Reduced-rank Vector Generalized Linear Models." *Statistical Modelling*, **3(1)**, 15–41.

Yee T, Wild C (1996). "Vector Generalized Additive Models." *Journal of the Royal Statistical Society B*, **58(3)**, 481–493.

Affiliation:

Benjamin Sexto Monroy
Department of Statistics, Colegio de Postgraduados
Campus Montecillo,Texcoco, Mexico 56230
E-mail: `bsexto@colpos.mx`

Humberto Vaquera Huerta
Department of Statistics, Colegio de Postgraduados,
Campus Montecillo, Texcoco, Mexico 56230
E-mail: `hvaquera@colpos.mx`

Barry C. Arnold
Department of Statistics, University of California, Riverside
Riverside, CA 92521
E-mail: `barry.arnold@ucr.edu`

11

Spatial Patterns of
Record-Setting Temperatures

Alex Kostinski
Department of Physics
Michigan Technological University

Amalia Anderson
Department of Physics
Hendrix College

Abstract

We employ record-breaking statistics to study spatial correlations of record-setting terrestrial surface temperatures. To that end, a simple diagnostic tool is devised, reminiscent of a pair-correlation function. Data analysis reveals that while during the hottest years, record-breaking temperatures arrive in "heat waves", extending throughout almost the entire continental United States, this is not so for all years, not even recently. Record-breaking temperatures generally exhibit spatial patterns and variability quite different from those of the mean temperatures.

Keywords: Record-setting, spatial correlation.

1. Introduction

Based on predictions of climate models, heat waves are widely expected to become more frequent and pronounced in the near future as climate change intensifies, e.g., Meehl and Tebaldi (2004). In the US, for example, last year was marked by extreme weather including droughts and record breaking heat in the middle west and Superstorm Sandy in the East. Thus, the perennial question on the public's mind is whether the anthropogenic climate change is responsible. This question of attribution is difficult as natural variability must be disentangled from the effects of a trend (i.e., global warming). Recent attempts to do so include work by Rahmstorf and Coumou (2011) who propose a statistical model that yields an increase of extreme events in the warming world. Most recently however, the mean global temperatures seem to have reached a plateau, e.g. see the recent perspective in Nature, entitled "The cause for the pause" (Held 2013). Perhaps one should pause for the cause instead. As a step toward disentangling natural variability from the possible trend, here we examine available evidence in terms of the "most extreme" variables, namely, record-breaking monthly temperatures within the continental United States.

Recent years have seen many applications of record-breaking to statistical physics and meteorology e.g. Wergen (2013); Edery, Kostinski, Majumdar, and Berkowitz (2013). Here our emphasis is on surface temperatures and here too, record-breaking statistics have been used, e.g. Benestad (2004); Meehl, Tebaldi, Walton, Easterling, and McDaniel (2009); Anderson and Kostinksi (2010). However, while the spatial distribution of records has been explored, e.g. Elguindi, Rauscher, and Giorgi (2012); Meehl, Arblaster, and Branstator (2012), to the best of our knowledge, spatial correlations in record-setting have not been examined quantitatively.

For the reader's convenience we briefly summarize essential definitions and mathematical results. The ith entry in a time-series, x_i, is a record-breaking event (record) if it exceeds all prior values in the sequence, that is, x_i is a record high if

$$x_i > \max(x_1, x_2, \ldots x_{i-1}) \tag{1}$$

and is a record low if

$$x_i < \min(x_1, x_2, \ldots x_{i-1}) \tag{2}$$

The first entry is always a record high and a record low by convention.

The crucial observation is the so-called reshuffling argument, namely, that for independent, identically distributed (i.i.d.), and continuous random variables, the nth trial has an equal chance of having the greatest value (denoted as $P_n(R)$) as all preceding trials, that is, $1/n$.

$$P_n(R) = 1/n \tag{3}$$

The expected number of records in a time series is the sum over trial probabilities of being a record. Thus the expected number of records, $E(R)$, for a time-series with n events is given by the harmonic series

$$E(R) = 1 + 1/2 + 1/3 \ldots + 1/n \tag{4}$$

and, by Euler's formula for large n

$$E(R) = \ln(n) + \gamma \tag{5}$$

where $\gamma = 0.577\ldots$, the Euler constant. These results are occasionally attributed to Rényi (1962) but in fact, originate with Foster and Stuart (1954). We stress the distribution independence of these results, i.e., they hold for any continuous probability densities. If the i.i.d. assumption is violated by a trend or by correlations, the number of records will deviate from the logarithmic dependence in Equation 5 and trends can, perhaps, be detected in a distribution-independent manner.

2. Problem Statement and Data Analysis

We shall now specialize the discussion, turning to terrestrial surface temperatures. Given the typical 100 years or so of temperature values, it becomes progressively less likely for any weather station to set a record high (or low), decaying as $1/n$ for stationary time series. In this sense, the classic lament by the elderly, e.g., "summers were hotter in my youth" is statistically valid insofar as record-breaking probabilities diminish as $1/n$ for stationary situations.

Global warming, of course, being a trend, works against the $1/n$ decay and much work has been done recently (some of it by us, Anderson and Kostinksi (2010)) to extract trends this way e.g. Benestad (2004); Redner and Petersen (2006); Wergen and Krug (2010); Rowe and Derry (2012). A brief consideration of the extreme cases of steep and weak trend limits will provide some perspective. In the steep trend limit (regardless of whether the trend is linear), every value is a record high and the expected number of records then scales as $E(R) \sim n^1$ while in the stationary case $E(R) \sim \ln(n)$ so one can expect the general behaviour $E(R) \sim n^\alpha$ with $0 < \alpha < 1$ but with a generally n-dependent pre-factor (Edery, Kostinski, and Berkowitz 2011).

The simplest description of a global warming trend buried in noise can be incorporated via the temperature time series of the form (Ballerini and Resnick 1985)

$$T_k = T_{sk} + ck \tag{6}$$

where T_{sk} is the stationary (random) value, k is an integer and c represents the "drift", characterizing global warming. Of course, T_k and, therefore, T_{sk} being random variables, are associated with some standard deviation (natural variability) σ and one expects the parameter $C \equiv c/\sigma$ to figure prominently in the modified record-breaking statistics (see, for example, Newman, Malamud, and Turcotte (2010) for a straightforward approach to C). Indeed, for the linear Gaussian drift model, for example, in the $C << 1$ approximation, one obtains (Franke, Wergen, and Krug 2010)

$$P_n \approx \frac{1}{n} + \frac{c}{\sigma} \frac{2\sqrt{\pi}}{e^2} \sqrt{\ln(\frac{n^2}{8\pi})} \tag{7}$$

Details of a specific model are not important for our purposes and the essential aspect is that while the drift c is viewed as "global", natural variability σ depends strongly on the location of a weather station as well as on the averaging period (e.g., daily vs. monthly temperatures). The latter is due to the range of temperature fluctuations e.g., being narrower near coasts, as well as by asymmetry of low and high temperatures, etc. A typical numerical value for the normalized drift c/σ, relevant to us below, is about 0.01 degrees per year (Wergen, Hense, and Krug 2013). However, as we noted earlier, the natural variability σ is not drawn from an ergodic ensemble because weather stations, say in coastal areas, differ greatly from those in the desert in Arizona. So the question is: given the the spatially heterogeneous natural variability, will the observed global drift alone suffice to set records in waves? Alternatively, can the normalized drift $C \equiv c/\sigma$ be regarded as a global parameter? If so, record-setting temperatures should arrive in bunches and wide-spread clusters on the entire globe. However, to the best of our knowledge, spatial correlations in record-setting have not been examined quantitatively.

A related question is: will record-breaking heat occur with spatial correlations similar to those of average monthly temperatures themselves? The answer is not obvious. Indeed, consider two nearby weather stations, say one in Chicago and another one in Milwaukee. Suppose that in 1921 (a relatively hot year), a heat wave reached and passed Chicago on the 17th of September but did not quite reach Milwaukee as the boundary of an associated front was sharper than the distance between the two cities. Hence, Chicago set a record high for that day but Milwaukee did not. Thus, the entire subsequent history of record-breaking in the two cities for all the 17 of September values were affected until such a time (if any) when a new record high occurred in both stations on Sept. 17. Clearly, although the underlying variables themselves (temperatures) are, likely, correlated because of the physical proximity of the two cities, it is not necessarily the case that the corresponding time-series of record-breaking events are similarly correlated.

It is well known that temperatures at nearby weather stations are correlated, e.g., the twin cities, St. Paul and Minneapolis, will have coinciding hot and cold years. Will the neighbouring stations also set records in unison? Correlation radius for monthly temperatures has already been studied rather thoroughly, for example Hansen and Lededeff (1987), who found that temperatures tend to be correlated up to a distance of 1200 km in mid to high latitudes (the correlation coefficient threshold was was set at 0.5). If temperatures are correlated out to 1200 km, are record breaking events similarly correlated? In fact, reflecting on the question leads one to realize that the very notion of the correlation coefficient is inadequate for the task at hand as record-breaking events depend heavily on prior history and, therefore, represent non-stationary time series even if the parent variable is a stationary one. To that end, below we propose a tool for characterizing such situations.

To address the above questions we use monthly mean temperatures from the United States Historical Climatology Network, version 2.5.0.20130501 (USHCN) (Menne, Williams, and Vose 2009). The US is chosen for this analysis because it is the densest region of stations extending back to 1900. Since our focus is on spatial correlation, a dense collection of stations is preferable. We use time series that have at least 90 years of data between the years 1900 and 2010 and expect 5.09 to 5.31 record-breaking events in a stationary and independent time series with 90 to 113 values. This results in 8290 time series (station-months) from about 690 stations (each month comprises its own time series); 38% of station-months were too short in duration and were excluded from the study. We use the adjusted data set, which accounts for irregularities in the raw data such as time of observation bias. Values that are estimates or are missing more than three entries in a monthly average are not used in this study, see Menne *et al.* (2009) and Menne, Williams, and Palecki (2010) for details regarding estimates and adjustments.

So, do record-setting temperatures arrive in spatial clusters and, if so, what is the cluster size? To answer these questions, we begin by examining the *fraction* of weather stations setting record highs and lows in the USHCN data. As a benchmark, we perform corresponding analyses for an independent, identically distributed (i.i.d.) Monte Carlo ensemble of the same dimensions as the used USHCN data, see Figure 1, panel (a). Immediately, the far greater variability of the fraction of stations with a record (y axis) in the USHCN data than the benchmark i.i.d. ensemble becomes evident. We see that the variability of both, record highs (orange) and record lows (blue) is much greater for USHCN than an i.i.d. Monte Carlo ensemble of the same dimensions. This contrast suggests that record-setting occurrences in USHCN time series are, indeed, correlated and, therefore cluster. Can the observed global warming trend (drift c) alone account for this? Panel (b) shows that the observed drift (mean trend, characterizing global warming) can not deliver such variability and, therefore, implicates spatial coherence of natural variability. The interplay is subtle, however, since the records certainly "know" about the mean trend: in panel (a) we see that between 1960 and

1980 there are several years with a deficiency of record highs and between 1998 and 2012 there is a deficiency of record lows. These are known cool and warm periods respectively, e.g. Shen, Lee, and Lawrimore (2012).

To test further, we mimic the mean trend in the USHCN data, e.g. Trenberth *et al.* (2007), and add a linear trend to an otherwise i.i.d. Monte Carlo ensemble (same dimensions are USHCN). We assume a normal distribution with an initial standard deviation equal to the average standard deviation for all series used in USHCN, $1.91°C$. The results are also shown in Figure 1, see panel (b). We use two trends mimicking those reported in IPCC for land in the Northern hemisphere: $\Delta\mu_1 = 0.063°C/\text{decade}$, 1850-2005, and $\Delta\mu_2 = 0.344°C/\text{decade}$, 1979-2005 (Trenberth *et al.* 2007). We see that with the simulated IPCC trends, the fraction of stations with a record has less variability than observed in the USHCN data and the fraction values are less extreme. Furthermore, i.i.d. Monte Carlo ensembles with piecewise trends have the same result – much less variability of fraction of highs or lows. Even when we compute a (LOWESS) smoothed trend based on USHCN anomalies and substitute it for a trend, we do not achieve the variability observed in the data. So, it is unlikely that the correlation, evident in the variability of results of Figure 1) is the result of the temporal trend alone. We now proceed to examine the likely spatial coherence in more detail.

To that end, we examine the behavior of records on either side of the boundary for correlation of the mean monthly temperatures themselves. Hansen and Lededeff (1987) found monthly temperatures to be correlated for about 1200 km (taking the threshold as correlation coefficient > 0.5). Therefore, we examined neighbouring stations in two categories: (1) neighbours with strongly correlated mean monthly temperatures (0-1200 km) and (2) neighbours with weakly correlated mean monthly temperatures (beyond 1200 km). As a benchmark, the corresponding analyses for i.i.d. Monte Carlo ensembles of the same dimensions as the used USHCN data are also shown in Figure 2. The analysis is as follows: for a single year, say 1950, for each record-breaking station, we compute the fraction of its neighbours (first nearer than 1200 km, panel (a), then farther than 1200 km, panel (b)) that also set temperature records. Then we compute the average of these fractions, to obtain the y-axis value in Figure 2. One expects that the closer stations are, the more likely they are to set records in unison. Indeed, in panel (a) we see, almost exclusively, an excess of records – when a station sets a record, its neighbours are more likely to set a record. This is true for both record highs and lows. Meanwhile, in panel (b) the message changes – distant neighbours do not always set records in unison, but instead may be more likely not to set a record. In fact, they are somewhat anti-correlated.

While Figure 2 suggests that 1200 km is a reasonable division between correlated and uncorrelated record-breaking, we want to look closer, considering the fractions of neighbours with records in concentric rings around a home station. Figure 3 demonstrates how we propose to do this. This is motivated by the notion of pair-correlation or radial distribution functions, often used in condensed matter physics. To find the fraction of records occurring as a function of distance from the base station at the origin, we consider concentric circles as depicted in panel (a). In panel (b), for each (base) station with a record-breaking high, we compute the fraction of its neighbours that also have a record high and then plot the average fraction. Note that for an extreme year like 2012 (the hottest in continental U.S. since 1900), the correlation

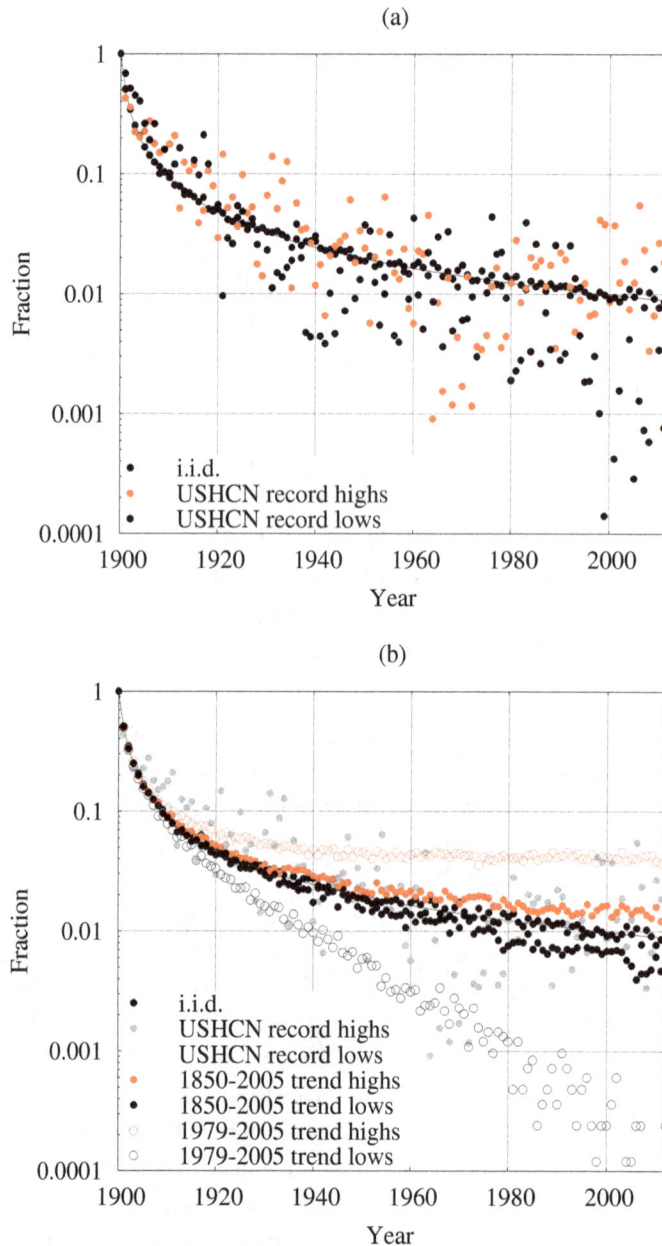

Figure 1: Fraction of stations setting records vs. the calendar year. Panel (a): the fraction of stations with a record high (orange) and low (blue) is shown per year for USHCN. Results for an i.i.d. Monte Carlo ensemble of the same dimensions as the used USHCN data is also shown (black). Panel (b): record breaking highs (orange) and lows (blue) for Monte Carlo ensembles of independent time series with two linear trends are shown that mimic trends reported by IPCC: 0.063°C/decade for 1850-2005 and 0.34°C/decade for 1979-2005 (Trenberth *et al.* 2007). For reference, results for USHCN record highs (gray filled circles) and lows (gray open circles) and an i.i.d. Monte Carlo ensemble (black) are included (same as in in panel (a)). The far greater variability of the fraction of stations with a record in the USHCN data versus the benchmark i.i.d. (identical and independently distributed) ensemble suggests that records in USHCN are correlated and, therefore, come in "bunches". Panel (b) shows that the observed mean trends alone do not account for the observed variability.

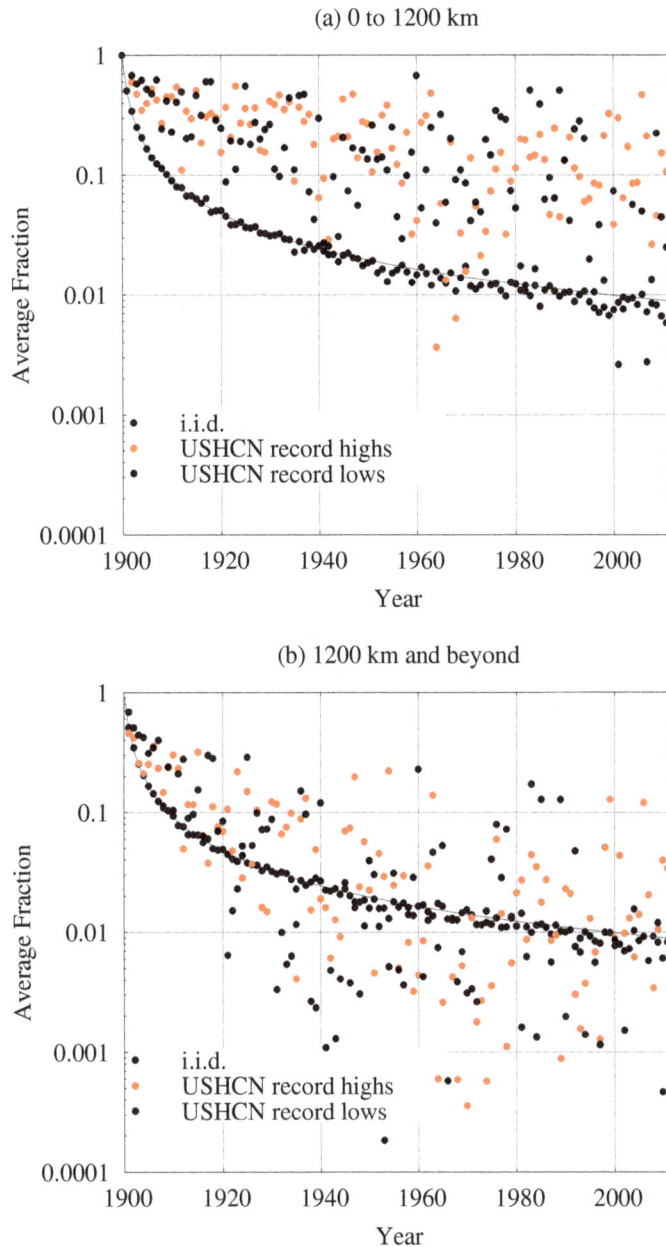

Figure 2: Same data as in the panel a of Figure 1 but separated into two parts: stations within 1200 km of each other (panel a) and the rest (figure b). The 1200 km boundary is chosen because it characterizes correlation radius for the monthly mean temperatures (see text). Record breaking highs and lows are spatially correlated out to at least 1200 km on average. The average fraction of neighbouring stations that set a record breaking high (orange) or low (blue) is shown for all stations that set a record breaking high. Neighbours in each panel are (panel a) those with correlated mean monthly temperatures (0-1200 km) and (panel b) stations with uncorrelated temperatures (1200 km and beyond); see text. Averages of the benchmark i.i.d. Monte Carlo ensembles are shown in black.

| | Hottest Years | | Coldest Years | | Median Years | |
Rank	Year	Relative Rank	Year	Relative Rank	Year	Relative Rank
1	2012	1 of 113	1917	18 of 18	1927	9 of 28
2	1998	1 of 99	1912	13 of 13	1942	18 of 43
3	2006	2 of 107	1924	23 of 25	1963	27 of 64
4	1931	1 of 32	1904	5 of 5	2008	53 of 109
5	1921	1 of 22	1978	75 of 79	2009	52 of 110

Table 1: In Figure 4 we examine records in three groups: 5 hot, 5 cold, and 5 typical years. The relative ranks for these years (hottest to coldest) are given here. The median years are listed chronologically rather in order of rank.

radius for record-setting temperatures far exceeds that of the monthly mean temperatures. Is this true for all years? Not so.

It turns out that the results of analysis illustrated in Figure 3 vary a great deal from year to year and this variation is summarized in Figure 4. We chose five hot, five cold and five unremarkable years to compute the correlation radius. These years are shown in Table 1 and Figure 4. The ranking is based on analysis of our USHCN subset (see above). Note, however, that other studies find slightly different rankings. For example, hottest years given by Shen et al. are 1998, 2006, 1934, 1921, 1999, etc. and the coldest years are 1917, 1895, 1912, 1924, 1903, etc., for the period 1985-2008, Shen *et al.* (2012).

It is evident from the panel (a) of Figure 4 that, during the hot years, spatial correlation extends well beyond 1200 km so that record-breaking heat arrives in huge waves. Somewhat surprisingly, in panel (b) we see similar pattern for record highs in one cold year (1904) but a much more rapid drop-off for other years. Overall, one expects fewer record highs in a cold year and this expectation is not in conflict with Figure 4. Observe also that if there is a record low set at the central station, its neighbours are likely to also set record lows. The correlation radius for these years appears to be well below the average for the five hottest years. Similarly, panel (c) shows that natural variability, as manifested by variation of the spatial correlation length from year to year, is remarkably pronounced. In passing, we note that Monte Carlo i.i.d. ensembles involve averaging over the number of stations with a record *and* with neighbours. The number of stations with records is constant along the x axis (approximately $1/n$, where n is number of years in time series, dictated by year), but the number of neighbours is not. The latter begins to drop off around 2500 km, and therefore, variability increases there (approximately $1/\sqrt{N}$, where N is the number of neighbours.)

The data are very interesting and somewhat puzzling. Why should the remarkably hot (hottest year on record) 2012 be almost matched in the spatial coherence of record-setting temperatures by the seemingly unremarkable 2009 (Figure 4, panel c)? Should the record-setting be attributable mostly to the global warming trend (drift) C in an otherwise random and spatially incoherent random field, the correlation radius would presumably follow the "heat" rank of the year. This does not appear to be the case and the underlying spatial coherence of the natural variability σ plays an important role.

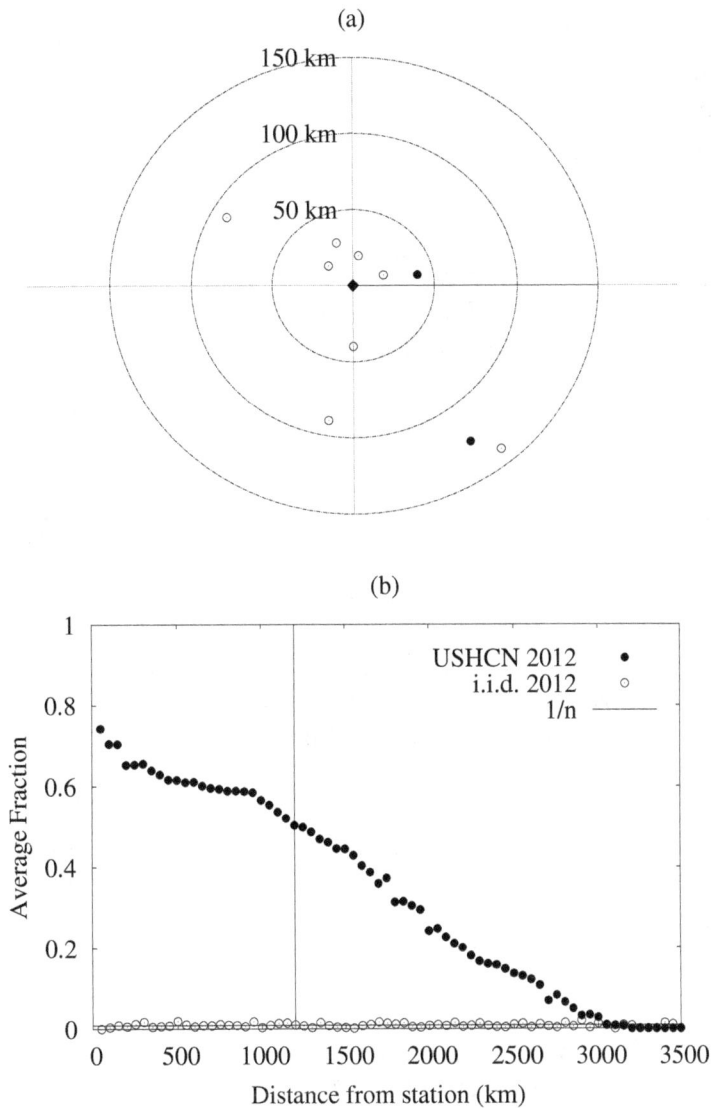

Figure 3: Average fraction of record-setting neighbours versus the distance from a given record-setting station. Panel (a): Computing the fraction of records occurring as a function of distance from a station at the origin, by counting within the concentric circles as depicted. A record-breaking station (black diamond) is at the origin and its neighbours are examined. Small circles represent stations, filled circles are stations with records. As illustrated in this figure, the fraction of record-breaking stations nearest to the station at the origin (within 0-50 km) is 1/6; 0/2 stations are between 50-100 km have a record and 1/2 stations between 100-150 km have a record. Panel (b): For each station with a record-breaking high, the fraction of its neighbours that also have a record high, is computed. The average fraction is depicted. A vertical line is shown at 1200 km, the point at which monthly temperatures drop below correlation coefficient of 0.5) (Hansen and Lededeff 1987). For an extreme year like 2012 (the hottest in continental U.S. since 1900), the correlation radius for records is considerably longer than anticipated by studies of the mean monthly temperatures.

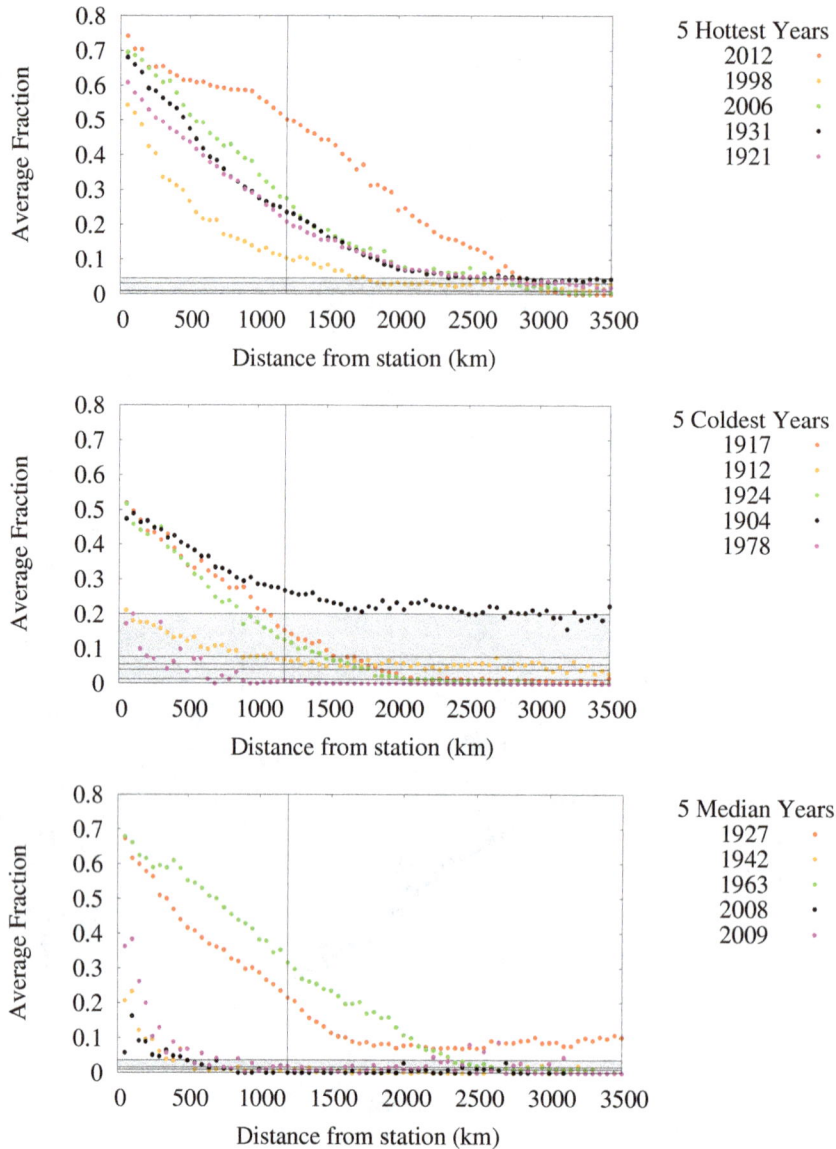

Figure 4: All panels: Average fraction of record-setting stations versus the distance from the base station. Panel (a): 5 hottest years. Panel (b); 5 coldest years. Panel (c): 5 typical years. 1900-2012. The average fraction of neighbouring stations that have a record breaking high is shown for all stations that have a record breaking high. A vertical line is shown at 1200 km, the point at which station temperature variables are considered to be uncorrelated (correlation defined as coefficient > 0.5) (Hansen and Lededeff 1987). Note that the correlation radius varies a great deal from year to year, indicating pronounced natural variability. I.i.d. Monte Carlo ensembles are included for reference, see the text for details. The grey shaded regions demonstrate the expected average fraction of neighbours with simultaneous record breaking. The upper limit corresponds to the shortest time series displayed, 22 years for (a), 5 years for (b), and 28 years for (c). Black horizontal lines below the upper limit correspond to expectations for longer time series corresponding again to the longer time periods shown.

3. Concluding Remarks

Viewing the attribution problem in terms of the competition between the global trend c and natural variability σ, the following question was posed here: given the the spatially heterogeneous natural variability, will the observed global drift alone suffice to set records in waves as observed? Alternatively, can the normalized drift $C \equiv c/\sigma$ be regarded as a global parameter? If so, record-setting temperatures should arrive in bunches and widespread clusters on the entire globe. However, spatial correlations in record-setting have not, to the best of our knowledge, been previously examined. Therefore, we used record-breaking statistics to examine spatial coherence of record-setting terrestrial surface temperatures, with the eye toward disentangling effects of global warming (mean drift or trend) from natural variability.

Devising an analogue of a pair-correlation function we have been able to examine record-setting temperatures by quantifying the size of underlying spatial clusters. We find that record-breaking of average monthly temperature time series exhibit spatial correlations quite different from those of the temperatures themselves. In particular, the variability of the correlation radius is much more pronounced for the record-breaking events. Furthermore, there is no clear correspondence between the extent of spatial correlations and the heat rank of the year, thereby indicating that the spatial coherence of natural variability plays an important role in setting up "waves" of extreme events.

Acknowledgments

This work was supported in part by the NSF grant AGS-111916.

References

Anderson A, Kostinksi A (2010). "Reversible record breaking and variability: Temperature distributions across the globe." *Journal of Applied Meteorology and Climatology*, **49**, 1681–1691.

Ballerini R, Resnick S (1985). "Records from improving populations." *Journal of Applied Probability*, **22**(3), 487–502.

Benestad RE (2004). "Record-values, nonstationarity tests and extreme value distributions." *Global and Planetary Change*, **44**(1-4), 11–26.

Edery Y, Kostinski AB, Berkowitz B (2011). "Record setting during dispersive transport in porous media." *Geophysical Research Letters*, **38**, L16403.

Edery Y, Kostinski AB, Majumdar SN, Berkowitz B (2013). "Record-breaking statistics for random walks in the presence of measurement error and noise." *Physical Review Letters*, **110**, 180602.

Elguindi N, Rauscher SA, Giorgi F (2012). "Historical and future changes in maximum and minimum temperature records over Europe." *Climatic Change*, **117**, 415–431.

Foster FG, Stuart A (1954). "Distribution-free tests in time-series based on the breaking of records." *Journal of the Royal Statistical Society: Series B*, **16**(1), 1–22.

Franke J, Wergen G, Krug J (2010). "Records and sequences of records from random variables with a linear trend." *Journal of Statistical Mechanics: Theory and Experiment*, **10**, P10013.

Hansen J, Lededeff S (1987). "Global trends of measured surface air temperature." *Journal of Geophysical Research*, **92**, D11 13,345–13,372.

Held IM (2013). "Climate science: The cause of the pause." *Nature*, **501**, 318–319.

Meehl GA, Arblaster JM, Branstator G (2012). "Mechanisms contributing to the warming hole and the consequent U.S. East-West differential of heat extremes." *Journal of Climate*, **25**, 6394–6408.

Meehl GA, Tebaldi C (2004). "More intense, more frequent, and longer laster heat waves in the 21st century." *Science*, **305**(5686), 994–997.

Meehl GA, Tebaldi C, Walton G, Easterling D, McDaniel L (2009). "Relative increase of record high maximum temperatures compared to record low minimum temperatures in the U.S." *Geophysical Research Letters*, **36**, L23701.

Menne MJ, Williams CN, Palecki MA (2010). "On the reliability of the U.S. surface temperature record." *Journal of Geophysical Research*, **115**, D11108.

Menne MJ, Williams CN, Vose RS (2009). "The United States Historical Climatology Network Monthly Temperature Data - Version 2." *Bulletin of American Meteorological Society*, **90**, 993–1007.

Newman WI, Malamud BD, Turcotte DL (2010). "Statistical properties of record-breaking temperatures." *Physical Review E*, **82**(6), 066111.

Rahmstorf S, Coumou D (2011). "Increase of extreme events in a warming world." *Proceedings of the National Academy of Sciences*, **108**(44), 17905–17909.

Redner S, Petersen MR (2006). "Role of global warming on the statistics of record-breaking temperatures." *Physical Review E*, **74**, 061114.

Rényi A (1962). "On the extreme elements of observations." In P Turán (ed.), *Selected Papers of Alréd Rényi*, volume 3. Akadémia Kiadó.

Rowe CM, Derry LE (2012). "Trends in record-breaking temperatures for the conterminous United States." *Geophysical Research Letters*, **39**(16), 6394.

Shen SSP, Lee CK, Lawrimore J (2012). "Uncertainties, trends, and hottest and coldest years of U.S. surface air temperature since 1895: An update based on the USHCN V2 TOB data." *Journal of Climate*, **25**, 4185–4203.

Trenberth KE, Jones PD, Ambenje P, Bojariu R, Easterling D, Klein Tank A, Parker D, Rahimzadeh F, Renwick JA, Rusticucci M, Soden B, Zhai P (2007). "Observations: Surface and Atmospheric Climate Change." In S Solomon, D Qin, M Manning, Z Chen, M Marquis, KB Averyt, M Tignor, HL Miller (eds.), *Climate Change 2007: The Physical Science Basis.*

Contribution of Working Group I to the Fourth Assessment Report of the Intergovernmental Panel on Climate Change. Cambridge University Press.

Wergen G (2013). "Records in stochastic processes - theory and applications." *Journal of Physics A: Mathematical and Theoretical*, **46**, 223001.

Wergen G, Hense A, Krug J (2013). "Record occurrence and record values in daily and monthly temperatures." *Climate Dynamics*, **42**(5-6), 1–15.

Wergen G, Krug J (2010). "Record-breaking temperatures reveal a warming climate." *EPL*, **92**, 30008.

Affiliation:

Alex Kostinksi
Department of Physics
Michigan Technological University
Houghton, MI 49931
E-mail: kostinsk@mtu.edu

Amalia Anderson
Department of Physics
Hendrix College
Conway, AR 72032
E-mail: andersona@hendrix.edu

12

Kernel Regression Model for Total Ozone Data

Horová I., Koláček J., Lajdová D.

Department of Mathematics and Statistics

Masaryk University Brno

Abstract

The present paper is focused on a fully nonparametric regression model for autocorrelation structure of errors in time series over total ozone data. We propose kernel methods which represent one of the most effective nonparametric methods.

But there is a serious difficulty connected with them – the choice of a smoothing parameter called a bandwidth. In the case of independent observations the literature on bandwidth selection methods is quite extensive. Nevertheless, if the observations are dependent, then classical bandwidth selectors have not always provided applicable results. There exist several possibilities for overcoming the effect of dependence on the bandwidth selection. In the present paper we use the results of Chu and Marron (1991) and Koláček (2008) and develop two methods for the bandwidth choice. We apply the above mentioned methods to the time series of ozone data obtained from the Vernadsky station in Antarctica. All discussed methods are implemented in Matlab.

Keywords: total ozone, kernel, bandwidth selection.

1. Introduction

Antarctica is significantly related to many environmental aspects and processes of the Earth. And thus its impact on the global climate system and water circulation in the world ocean is essential.

The stratosphere ozone depletion over Antarctica was discovered at the beginning of the 1990s. The lowest total ozone contents (TOC) in Antarctica are usually observed in the first week of October. The formation of ozone depletion begins approximately in the second half of August, culminates in the first half of October, and dissolves in November. During the ozone depletion, the average ozone concentration varied at the time of its culmination in October from the original value over 300 Dobson Units (DU) in 1950s and 1960s to a level between 100 and 150 DU in 1990-2000 (see Láska *et al.* (2009)). One DU is set as a 0.001 mm strong layer of ozone under the pressure 1013 hPa and temperature 273 K.

One of the issues resolved within the Czech–Ukrainian scientific cooperation implemented on the Vernadsky Station in Antarctica is the measurement of total ozone content (TOC) in the stratosphere. The Vernadsky station is located on the west coast of Antarctic peninsula (65°S, 64°W). These data were obtained from ground measurements predominantly taken with the Dobson No 031 spectrophotometer. Data can be found at UAC (2012).

The data sets were processed as time points measuring the average daily amount of ozone. In order to analyze these data we have to take into account the autocorrelation structure of errors on such time series. We focus on kernel regression estimators of series of ozone data. These estimators depend on a smoothing parameter and it is well-known that selecting the correct smoothing parameter is difficult in the presence of correlated errors. There exist methods which are modifications of a classical cross-validation method for independent errors (the modified cross-validation method or the partitioned cross-validation method - see Chu and Marron (1991), Härdle and Vieu (1992)).

In the present paper we develop a new flexible plug-in approach for estimating the optimal smoothing parameter. The utility of this method is illustrated through a simulation study and application to TOC data measured in periods August to April 2004-2005, 2005-2006, 2006-2007.

2. Procedure Development

2.1. Kernel regression model

In nonparametric regression problems we are interested in estimating the mean function $E(Y|x) = m(x)$ from a set of observations (x_i, Y_i), $i = 1, \ldots, n$. Many methods such as kernel methods, regression splines and wavelet methods are currently available. The papers in this filed have been mostly focused on case where an unknown function m is hidden by a certain amount of a white noise. The aim of a regression analysis is to remove the white noise and produce a reasonable approximation to the unknown function m.

Consider now the case when the noise is no longer white and instead contains a certain amount of a structure in the form of correlation. In particular, if data sets have been recorded over time from one object under a study, it is very likely that another response of the object will depend on its previous response. In this context we will be dealing with a time series case, where design points are fixed and equally spaced and thus our model takes the form

$$Y_i = m(i/n) + \varepsilon_i, \quad i = 1, \ldots, n, \tag{1}$$

and ε_i is an unknown ARMA process, i.e.,

$$\begin{aligned} E(\varepsilon_i) &= 0, \quad \mathrm{var}(\varepsilon_i) = \sigma^2, \quad i = 1, \ldots, n, \\ \mathrm{cov}(\varepsilon_i, \varepsilon_j) &= \gamma_{|i-j|} = \sigma^2 \rho_{|i-j|}, \quad \mathrm{corr}(\varepsilon_i, \varepsilon_j) = \rho_{|i-j|} \end{aligned} \tag{2}$$

and the stationary process

$$\gamma_0 = \sigma^2, \quad \rho_t = \frac{\gamma_t}{\gamma_0},$$

where ρ_t is an autocorrelation function and γ_t is an autocovariance function. We consider the simplest situation (Opsomer *et al.* (2001), Chu and Marron (1991))

$$\rho_{t/n} = \rho_t.$$

Simple and the most widely used regression smoothers are based on kernel methods (see e.g. monographs Müller (1987), Härdle (1990), Wand and Jones (1995)). These methods are local weighted averages of the response Y. They depend on a kernel which plays the role of a weighted function, and a smoothing parameter called a bandwidth which controls the smoothness of the estimate.

Appropriate kernel regression estimators were proposed by Priestley and Chao (1972), Nadaraya (1964) and Watson (1964), Stone (1977), Cleveland (1979) and Gasser and Müller (1979).

These estimators were shown to be asymptotically equivalent (Lejeune (1985), Müller (1987), Wand and Jones (1995)) and without the lost of generality we consider the Nadaraya–Watson (NW) estimators $\widehat{m}$ of m. The NW estimator of m at the point $x \in (0,1)$ is defined as

$$\widehat{m}(x,h) = \frac{\sum\limits_{i=1}^{n} K_h(x_i - x)Y_i}{\sum\limits_{i=1}^{n} K_h(x_i - x)}, \tag{3}$$

for a kernel function K, where $K_h(.) = \frac{1}{h}K(\frac{\cdot}{h})$, and h is a nonrandom positive number $h = h(n)$ called the bandwidth.

Before studying the statistical properties of $\widehat{m}$ several additional assumptions on the statistical model and the parameters of the estimator are needed:

I. Let $m \in C^2[0,1]$.

II. Let K be a real valued function continuous on $\mathbb{R}$ and satisfying the conditions:

   (i) $|K(x) - K(y)| \le L|x - y|$ for a constant $L > 0$, $\forall x, y \in [-1,1]$,

   (ii) $\text{support}(K) = [-1,1]$, $K(-1) = K(1) = 0$,

   (iii) $\int_{-1}^{1} x^j K(x)\mathrm{d}x = \begin{cases} 1 & j = 0, \\ 0 & j = 1, \\ \beta_2 \neq 0 & j = 2. \end{cases}$

   Such a function is called a kernel of order 2 and a class of these kernels is denoted as S_{02}.

III. Let $h = h(n)$ be a sequence of nonrandom positive numbers, such that $h \to 0$ and $nh \to \infty$ as $n \to \infty$.

IV. $\lim\limits_{n \to \infty} \sum\limits_{k=1}^{\infty} |\rho_k| < \infty$,   i.e., $R = \sum\limits_{k=1}^{\infty} \rho_k$ exists,

V. $\frac{1}{n} \sum\limits_{k=1}^{\infty} k|\rho_k| = 0$.

Remark. The well-known kernels are, e.g.,

Epanechnikov kernel $K(x) = \frac{3}{4}(1 - x^2)I_{[-1,1]}$,

quartic kernel $K(x) = \frac{3}{4}(1 - x^2)^2 I_{[-1,1]}$,

triweight kernel $K(x) = \frac{35}{32}(1 - x^2)^2 I_{[-1,1]}$,

Gaussian kernel $K(x) = \frac{1}{\sqrt{2\pi}}e^{\frac{-x^2}{2}}$,

where $I_{[-1,1]}$ is an indicator function.

Though the Gaussian kernel does not satisfy the assumption II.(ii), it is very popular in many applications.

There is no problem with a choice of a suitable kernel. Symmetric probability density functions are commonly used (see Remark above). But choosing the smoothing parameter is a crucial problem in all kernel estimates. The literature on bandwidth selections is quite extensive in case of independent errors.

It is well known that when the kernel method is used to recover m, that correlated errors trouble bandwidth selection severely (see Altman (1990), Opsomer *et al.* (2001)). De Brabanter *et al.* (2010) developed a bandwidth selection procedure based on bimodal kernels which successfully removes the error correlation without requiring any prior knowledge about its structure.

The global quality of the estimate $\widehat{m}$ can be expressed by means of the Mean Integrated Squared Error (Altman (1990), Opsomer *et al.* (2001)). However more mathematically tractable is the Asymptotic Mean Integrated Squared Error (AMISE):

$$\text{AMISE}(\widehat{m}, h) = \underbrace{\frac{V(K)}{nh}S}_{\text{AIV}(\widehat{m},h)} + \underbrace{\frac{\beta_2^2}{4}h^4 A_2}_{\text{AISB}(\widehat{m},h)},$$

where

$$V(K) = \int K^2(x)\mathrm{d}x, \quad S = \sigma^2\left(1 + 2\sum_{k=1}^{\infty} \rho_k\right) = \sigma^2(1 + 2R), \quad A_2 = \int_0^1 m''(x)^2\mathrm{d}x.$$

The first term is called the asymptotic integrated variance (AIV) and the second one the asymptotic integrated squared bias (AISB). This decomposition provides an easier analysis and interpretation of the performance of the kernel regression estimator.

Using a standard procedure of mathematical analysis one can easily find that the bandwidth h_{opt} minimizing the AMISE is given by the formula

$$h_{opt} = \left(\frac{V(K)S}{n\beta_2^2 A_2}\right)^{1/5} = O(n^{-1/5}). \tag{4}$$

This formula provides a good insight into an optimal bandwidth, but unfortunately it depends on the unknown S and A_2.

Let us explain the impact of assuming an uncorrelated model.

If $R > 0$ (error correlation is positive), then $\mathrm{AIV}(\widehat{m}, h)$ is larger than in the corresponding uncorrelated case and $\mathrm{AMISE}(\widehat{m}, h)$ is minimized by a value h that is larger than in the uncorrelated case. It means that assuming wrongly uncorrelated errors causes that the bandwidth becomes too small.

If $R < 0$ (error correlation is negative), then $\mathrm{AIV}(\widehat{m}, h)$ is smaller and $\mathrm{AMISE}(\widehat{m}, h)$ optimal bandwidth is smaller than in the uncorrelated case.

In the next section the choosing of parameters S and A_2 will be treated.

2.2. Choosing the parameters

There are a number of data-driven bandwidth selection methods, but it can be shown that they fail in the case of correlated errors.

Among the earliest fully automatic and consistent bandwidth selectors are those based on cross-validation ideas. The cross-validation method employs an objective function

$$CV(h) = \frac{1}{n} \sum_{j=1}^{n} \left(\widehat{m}_{-j}(x_j, h) - Y_j \right)^2, \tag{5}$$

where $\widehat{m}_{-j}(x_j, h)$ is the estimate of $\widehat{m}(x_j, h)$ with x_j deleted, i.e., the leave-one-out estimator. The estimate of h_{opt} is then

$$\widehat{h}_{opt} = \arg \min_{h \in H_n} CV(h),$$

where $H_n = [an^{-1/5}, bn^{-1/5}], \quad 0 < a < b < \infty.$

Remark. If the design points are equally spaced then a recommended interval is $[\frac{1}{n}, 1)$.

However, this ordinary method is not suitable in the case of correlated observations. As it was shown in the papers Altman (1990) and Opsomer *et al.* (2001), if the observations are positively correlated, then the CV method produces too small a bandwidth, and if the observations are negatively correlated, then the CV method produces a large bandwidth.

We demonstrate this fact by the following example.

Consider the regression model (1), where

$$m(x) = \cos(3.15\pi x), \quad \varepsilon_i = \phi\varepsilon_{i-1} + e_i,$$

$\qquad e_i$ – i.i.d. normal random variables $\mathrm{N}(0, \sigma^2)$,

$\qquad \varepsilon_1 - \mathrm{N}(0, \sigma^2/(1 - \phi^2))$,

$\qquad \phi = 0.6, \quad \sigma = 0.5,$

$\qquad$ i.e, the regression errors are AR(1) process.

Figure 1 shows the result obtained by the CV method. It is evident, that the estimate is undersmoothed.

In order to overcome this problem, modified and partitioned CV methods were proposed by Härdle and Vieu (1992) and Chu and Marron (1991), respectively.

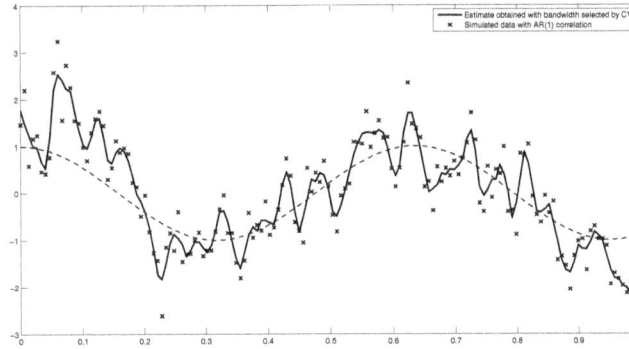

Figure 1: The estimate of simulated data with AR(1) errors

The modified cross-validation (MCV) method is a "leave-$(2l + 1)$-out" version of CV ($l \geq 0$). The idea consists in minimizing of the modified cross-validation score:

$$CV_l(h) = \frac{1}{n} \sum_{j=1}^{n} \left(\widehat{m}_{-j}(x_j, h) - Y_j \right)^2, \tag{6}$$

where $\widehat{m}_{-j}(x_j, h)$ is the "leave-$(2l+1)$-out" estimate of $\widehat{m}(x_j, h)$, i.e., the observations (x_{j+i}, Y_{j+i}), $-l \leq i \leq l$ are left out in constructing $\widehat{m}(x_j, h)$.

Then

$$\widehat{h}_{MCV} = \arg \min_{h \in H_n} CV_l(h).$$

The principle of the partitioned cross-validation method (PCV) can be described as follows.

For any natural number $g \geq 1$, the PCV involves splitting the observations into g groups by taking every g-th observation, calculating the ordinary cross-validation score $CV_{0,k}(h)$ of the k-th group of observations separately, for $k = 1, 2, \ldots, g$, and minimizing the average of these ordinary cross-validation scores

$$CV^*(h) = \frac{1}{g} \sum_{k=1}^{g} CV_{0,k}(h). \tag{7}$$

Let $\widehat{h}_{CV}^*$ stand for the minimizer of $CV^*(h)$:

$$\widehat{h}_{CV}^* = \arg \min_{h \in H_n} CV^*(h).$$

Since $\widehat{h}_{CV}^*$ is appropriate for the sample size n/g, the partitioned cross-validated bandwidth $\widehat{h}_{PCV(g)}$ is defined to be rescaled $\widehat{h}_{CV}^*$:

$$\widehat{h}_{PCV(g)} = g^{-1/5} \widehat{h}_{CV}^*.$$

When $g = 1$, the PCV is an ordinary cross-validation.

Remark. The number of subgroups is g and the number of observations in each group is $\eta = n/g$. If n is not a multiplier of g, then the values Y_j, $1 \leq j \leq g[n/g]$ are applied and the rest of the observations are dropped out ($[n/g]$ is the highest integer less or equal to n/g).

The asymptotic behavior of $\widehat{h}_{MCV(l)}$ and $\widehat{h}_{PCV(g)}$ was studied in the paper by Chu and Marron (1991). Furthemore we focus on the PCV method.

The PCV method needs to determine the factor g. A possible approach for the practical choice of g is based on an analogue of the mean squared error. Using the asymptotic variance and the asymptotic mean of $\widehat{h}_{PCV(g)}/h_{opt}$, the asymptotic mean squared error (AMSE) of this ratio is defined by

$$AMSE\big(\widehat{h}_{PCV(g)}/h_{opt}\big) = n^{-1/5}\text{VAR}_{PCV(g)} + \big[C_{PCV(g)}/C - 1\big]^2, \tag{8}$$

where $\text{VAR}_{PCV(g)}$, $C_{PCV(g)}$, C depend on γ_k, K, A_2 (see Chu and Marron (1991)).

Theoretically, if there exists a value $\widehat{g}$ which minimizes AMSE over $g \geq 1$, then this value is taken as the optimal value of g in the sense of AMSE:

$$\widehat{g}_{opt} = \arg\min_{g \geq 1} \text{AMSE}\big(\widehat{h}_{PCV(g)}/h_{opt}\big).$$

Unfortunately the minimization of AMSE also depends on the unknown γ_k and A_2.

As far as the estimation of the variance component S is concerned, a common approach is the following (see e.g. Herrmann *et al.* (1992), Hart (1991), Opsomer *et al.* (2001), Chu and Marron (1991)):

$$\widehat{S} = \widehat{\gamma}_0\Big(1 + 2\sum_{k=1}^{n-1}\widehat{\rho}_k\Big), \quad \widehat{\gamma}_0 = \widehat{\sigma}^2, \quad \widehat{\rho}_k = \frac{\widehat{\gamma}_k}{\widehat{\gamma}_0},$$

$$\widehat{\gamma}_k = \frac{1}{n-k}\sum_{t=1}^{n-k}\Big(Y_t - \overline{Y}\Big)\Big(Y_{t+k} - \overline{Y}\Big), \quad k = 0,\ldots,n-1. \tag{9}$$

Nevertheless there is still a problem of how to estimate A_2. In paper Chu and Marron (1991) a simulation study was only conducted and no idea of estimating A_2 was given there.

We complete this method by adding a suitable estimate of A_2 and recommend to use an estimate of A_2 proposed by Koláček (2008). By means of the Fourier transformation he derived a suitable estimate $\widehat{A_2}$ of A_2. Therefore, A_2 in the AMSE formula is replaced by $\widehat{A_2}$. This approach is commonly known as a plug-in method.

Plug-in methods are also commonly used for selecting the bandwidth in the kernel regression. But these methods perform badly when the errors are correlated. In the paper Herrmann *et al.* (1992) a modified version of an existing plug-in bandwidth selectors is proposed. This method is based on the Gasser–Müller estimator of the second derivative and an iterative process is constructed. It is shown that under some additional assumptions this iterative process converges to a suitable estimate of the optimal bandwidth.

However we do not use this iterative method and propose to directly plug-in A_2 in the formula (4). This new version of a plug-in method is denoted as PI and the bandwidth estimate takes the form:

$$\widehat{h}_{PI} = \Big(\frac{V(K)\widehat{S}}{n\beta_2^2\widehat{A_2}}\Big)^{1/5}.$$

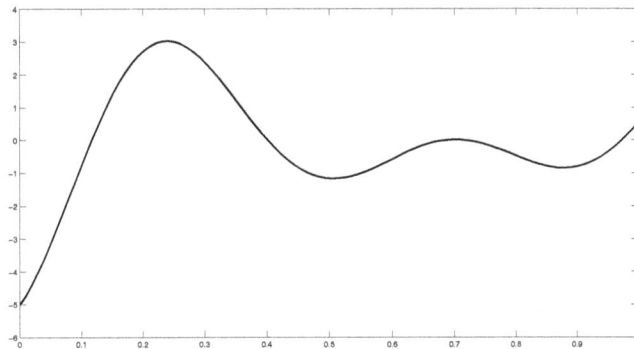

Figure 2: The regression function $m(x)$

$h_{opt} = 0.759$		
	$E(\widehat{h})$	$std(\widehat{h})$
PCV	0.1927	0.0649
PI	0.1513	0.0083

Table 1: The estimates $\widehat{h}$

We would like to point out the computational aspect of the plug-in method. It has preferable properties to classical methods, because it does not need any additional calculations such as the PCV method (see Koláček (2008) for details).

3. Case study

We conduct a simulation study to compare the PCV method and the PI method. The Epanechnikov kernel is used both in simulations and in applications.

Consider the regression model (1), where

$$m(x) = \frac{-6\sin 11x + 5}{\cotg(x-7)}, \quad \varepsilon_i = \phi\varepsilon_{i-1} + e_i$$

e_i – i.i.d. normal random variables $N(0, \sigma^2)$
$\varepsilon_1 - N(0, \sigma^2/(1-\phi^2))$
$\phi = 0.6, \quad \sigma = 0.5,$

for $i = 1, \ldots, n = 100$.

The graph of the regression function m is presented in Figure 2.

One hundred series are generated. For each data set, the optimal bandwidth is estimated by the PCV and PI method. Table 1 shows the comparison of means and standard deviations for these two methods.

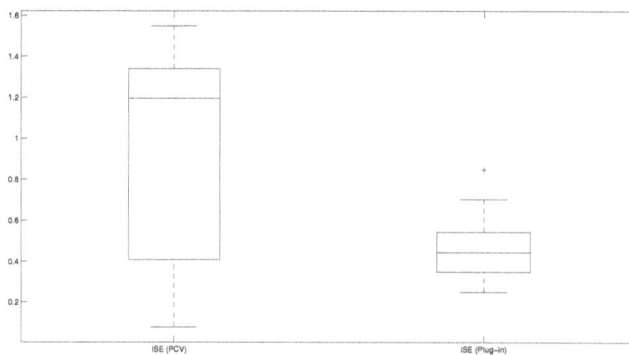

Figure 3: $ISE(\widehat{m}(.,h)) = \int_0^1 \Big(\widehat{m}(x,h) - m(x)\Big)^2 dx.$

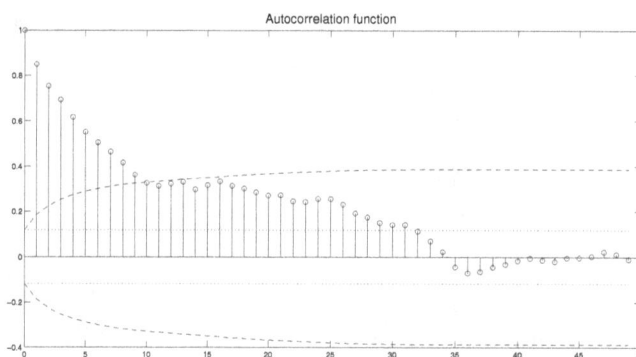

Figure 4: The autocorrelation function of the data set August 2004 – April 2005

The Integrated Square Error (ISE) is calculated for each estimate $\widehat{m}(.,h)$:

$$ISE(\widehat{m}(.,h)) = \int_0^1 \Big(\widehat{m}(x,h) - m(x)\Big)^2 dx$$

for both PCV and PI methods and the results are displayed by means of the boxplots in Figure 3.

4. Results and discussion

In this section we apply the methods described above to ozone data. We analyze data which were measured in the period August to April in years 2004–2005, 2005–2006, 2006–2007. The sample size is $n = 273$ days. The observations are correlated as it can be seen in Figure 4. We transform data to the interval [0,1] and use the PCV method and the PI method to get the optimal bandwidth. Then we re-transform the bandwidth to the original sample and obtain the final kernel estimate.

Kernel estimates based on the PCV and PI methods are presented in Figure 6, Figure 7, or in Figure 8, respectively.

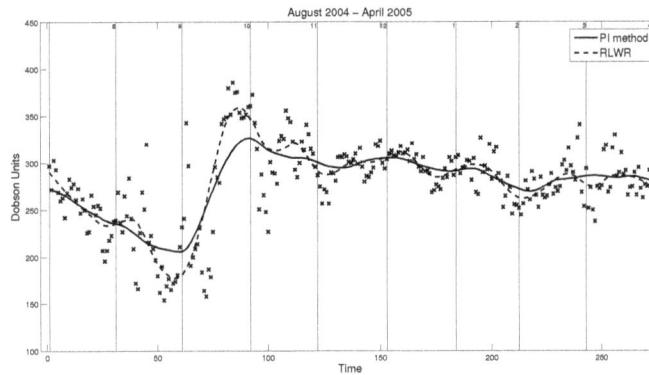

Figure 5: RLWR estimate with span = 40 (dashed line) and PI estimate with the bandwidth = 17.8 (solid line).

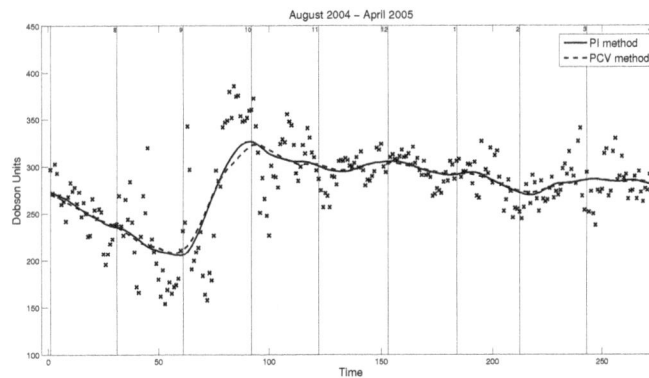

Figure 6: PCV estimate with the bandwidth = 20.9 (dashed line) and PI estimate with the bandwidth = 17.8 (solid line).

In paper Kalvová and Dubrovský (1995) the robust locally wighted regression (RLWR) is employed for data processing of TOC. They recommended to optimize h subjectively. This approach needs an experience and a special knowledge of the given data sets. The advantage of our methods consists in more complex approach. These methods are general and they allow to choose the value of h automatically. We used their methodology for data April 2004 - August 2005 and the comparison of the estimate obtained by the PI method and by the robust locally weighted regression can be seen in Figure 5. The PI method yields a rather oversmoothed estimate.

Our experience shows that both methods could be considered as a suitable tool for the choice of the bandwidth. But it seems that the PI method is sufficiently reliable and less time consuming than the PCV method.

Presented methods can be applied to other time series not only in environmetrics but also in economics or other fields.

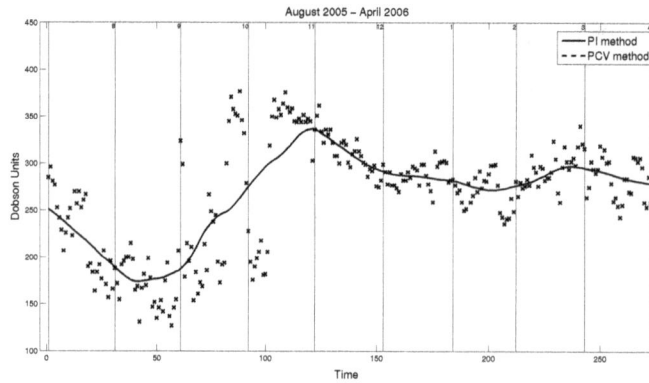

Figure 7: PCV estimate with the bandwidth = 20.4 (dashed line) and PI estimate with the bandwidth = 21.9 (solid line).

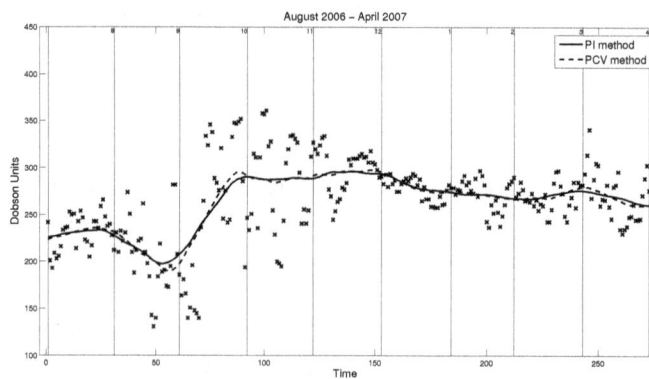

Figure 8: PCV estimate with the bandwidth = 17.2 (dashed line) and PI estimate with the bandwidth = 22.3 (solid line).

Acknowledgments

The research was supported by The Jaroslav Hájek Center for Theoretical and Applied Statistics (MŠMT LC 06024). The work was supported by the Student Project Grant at Masaryk university, rector's programme no. MUNI/A/1001/2009.

References

Altman N (1990). "Kernel Smoothing of Data With Correlated Errors." *Journal of the American Statistical Association*, **85**, 749–759.

Chu CK, Marron JS (1991). "Choosing a Kernel Regression Estimator." *Statistical Science*, **6**(4), 404–419. ISSN 08834237.

Cleveland WS (1979). "Robust Locally Weighted Regression and Smoothing Scatterplots." *Journal of the American Statistical Association*, **74**(368), 829–836. ISSN 01621459.

De Brabanter K, De Brabanter J, Suykens J, De Moor B (2010). "Kernel Regression with Correlated Errors." *Computer Applications in Biotechnology*, pp. 13–18.

Gasser T, Müller HG (1979). "Kernel estimation of regression functions." In T Gasser, M Rosenblatt (eds.), *Smoothing Techniques for Curve Estimation*, volume 757 of *Lecture Notes in Mathematics*, pp. 23–68. Springer Berlin / Heidelberg.

Härdle W (1990). *Applied Nonparametric Regression*. 1st edition. Cambridge University Press, Cambridge.

Härdle W, Vieu P (1992). "Kernel Regression Smoothing of Time Series." *Journal of Time Series Analysis*, **13**(3), 209–232.

Hart JD (1991). "Kernel Regression Estimation with Time Series Errors." *Journal of the Royal Statistical Society*, **53**, 173–187.

Herrmann E, Gasser T, Kneip A (1992). "Choice of Bandwidth for Kernel Regression when Residuals are Correlated." *Biometrika*, **79**, 783–795.

Kalvová J, Dubrovský M (1995). "Assessment of the Limits Between Which Daily Average Values of Total Ozone Can Normally Vary." *Meteorol. Bulletin*, **48**, 9–17.

Koláček J (2008). "Plug-in Method for Nonparametric Regression." *Computational Statistics*, **23**(1), 63–78. ISSN 0943-4062.

Láska K, Prošek P, Budík L, Budíková M, Milinevsky G (2009). "Prediction of Erythemally Effective UVB Radiation by Means of Nonlinear Regression Model." *Environmetrics*, **20**(6), 633–646.

Lejeune M (1985). "Estimation Non-paramétrique par Noyaux: Régression Polynomiale Mobile." *Revue de Statistique Appliquée*, **33**(3), 43–67.

Müller HG (1987). "Weighted Local Regression and Kernel Methods for Nonparametric Curve Fitting." *Journal of the American Statistical Association*, **82**(397), 231–238. ISSN 01621459.

Nadaraya EA (1964). "On Estimating Regression." *Theory of Probability and its Applications*, **9**(1), 141–142.

Opsomer J, Wang Y, Yang Y (2001). "Nonparametric Regression with Correlated Errors." *Statistical Science*, **16**(2), 134–153.

Priestley MB, Chao MT (1972). "Non-Parametric Function Fitting." *Journal of the Royal Statistical Society. Series B (Methodological)*, **34**(3), 385–392. ISSN 00359246.

Stone CJ (1977). "Consistent Nonparametric Regression." *The Annals of Statistics*, **5**(4), 595–620. ISSN 00905364.

UAC (2012). "World Ozone and Ultraviolet Radiation Data Centre (WOUDC) [data]." URL http://www.woudc.org.

Wand M, Jones M (1995). *Kernel smoothing*. Chapman and Hall, London.

Watson GS (1964). "Smooth Regression Analysis." *Sankhya - The Indian Journal of Statistics, Series A*, **26**(4), 359–372. ISSN 0581572X.

Affiliation:

Ivana Horová
Masaryk University
Department of Mathematics and Statistics
Brno, Czech Republic
E-mail: horova@math.muni.cz

13

Filtering and functional parameter estimation of spatiotemporal strong-dependence models

María Pilar Frías
Department of Statistics
and Operations Research,
University of Jaén, Spain

María Dolores Ruiz-Medina
Department of Statistics
and Operations Research,
University of Granada, Spain

Abstract

Filtering and parameter estimation are addressed in the context of spatiotemporal strong dependence processes. A functional parametric observation model is fitted to the spectral sample information. Specifically, the class of strong dependence spatiotemporal random fields studied in Frías, Ruiz-Medina, Alonso, and Angulo (2006a), Frías, Ruiz-Medina, Alonso, and Angulo (2008), Frías, Ruiz-Medina, Alonso, and Angulo (2009) is considered. Large dimensional spectral data sets, displaying high local singularity, are then processed in this functional setting. Thresholding techniques are first applied for removing noise generated from measurement spectrometer device. Spatiotemporal long-range dependence model fitting is then achieved by applying linear regression in the log-wavelet domain. The performance of the estimation algorithms proposed is illustrated from simulated data.

Keywords: Fractal spectral processes, long-range dependence parameters, spatiotemporal parametric models, wavelet thresholded transform.

1. Introduction

Remote Sensing of Environment constitutes a crucial task in assessment of urban heat island effect, surface soil water content, green vegetation, climate change, etc. (see Byambakhuu, Sugita, and Matsushima (2010); Elvidge, Chen, and Groeneveld (1993); Gallo, McNab, Karl, Brown, Hood, and Tarpley (1993); Price (1990), among others). Imaging spectrometry is one of the most common tools used for capturing sample information from the earth (see, for example, Bell, B.A., and Martini (2010); Clark and Roush (1984); Goetz, Vane, Solomon, and Rock (1985)). This paper contributes to the analysis of spatiotemporal data in this context, since functional data are collected in the spectral domain, focusing our study to the case of strong correlations in space and time. That is, the collected functional spectral data displays a singularity at the origin (see, Leonenko (1999) and Frías *et al.* (2009)), considering the statistical functional framework (see, Delicado, Giraldo, Comas, and Mateu (2009)).

In the spatial statistical analysis of high dimensional data sets, for example, in the investigation of soil properties, air ozone concentration, velocity turbulence fields, ocean surface temperature profiles from deep ocean weather stations, often strong correlations are detected in time and or space (see, for example, Akkaya and Yücemen (2002), Anh, Lam, Leung, and Tieng (2000), Marguerit, Schertzed, Schmitt, and Lovenjoy (1998)). For the analysis of high-dimensional data usually dimension reduction techniques are applied. In the strong-dependence case these techniques must be combined with the truncation of the heavy tails of the underlying covariance model. In particular, this fact corresponds to the truncation of the singularity at zero of the associated spectral density, which provides information on the behavior of the covariance function at large lags. The hyperbolic rate displayed by the decay velocity of the covariance function of strong-correlated random fields induces serious difficulties in the implementation of inference tools (see, Frías *et al.* (2009)). This fact hinders the development of functional estimation algorithms, since several model selection and matrix computational problems arise. Some attempts have been made in the parametric and semi-parametric spatial and spatiotemporal estimation contexts (see Bardet, Lang, Oppenheim, Philippe, Stove, and Taqqu (2003), Frías *et al.* (2006a), Frías *et al.* (2008), Frías *et al.* (2009), among others). A complete overview on statistical inference tools for random fields with singular spectra can be found in Leonenko (1999) (see also Kelbert, Leonenko, and Ruiz-Medina (2005)).

The measurement noise associated with devices, like spectrometers, increases the high local singular nature of spectral data, whose covariance function is heavy-tailed. This fact motivates the methodology proposed in this paper for addressing the problem of spatiotemporal long-range dependence model fitting from noisily large-dimensional spectral data. Specifically, the wavelet domain is first considered to transform the large-dimensional spectral data into functional data. Thresholding techniques are applied to the empirical wavelet coefficients for removing the observation noise (see, for example, Donoho and Johnstone (1995)). Linear regression estimates are then computed from the log-thresholded wavelet transform to approximate the temporal and spatial long-range dependence parameters.

The estimation approach presented in this paper can also be applied to the functional non-parametric regression context in the case where the functional regressors display strong correlation (see Ferraty, Goia, and Vieu (2002), under the weak-dependence modeling, and Benhenni, Hedli-Griche, Rachdia, and Vieu (2008), under long memory conditions). In the Geostatistical framework, spatiotemporal kriging and functional parameter estimation, under strong dependence modeling, can also be addressed from the application of the estimation methodology formulated here. In this context, we mention the papers by Delicado *et al.* (2009), Baladandayuthapani, Mallick, Hong, Lupton, Turner, and Caroll (2008) and Basse, Diop, and Dabo-Niang (2008) on statistical analysis of functional data displaying spatial interaction (see also Ruiz-Medina (2011), in the spatial autoregressive functional context).

The outline of the paper is the following. The spatiotemporal strong dependence functional model, assumed in the development of the results presented in this paper, is introduced in Section 2. The filtering and parameter estimation methodology proposed are described in Section 3. A simulation study is carried out in Section 4 for illustration of the wavelet-based filtering and long-range dependence model fitting approach presented in the spectral domain. A real-data example is considered in Section 5 to illustrate the methodology proposed. Concluding remarks are given in Section 6.

2. Statistical functional model

Let us consider the following functional spectral observation model:

$$Z(\omega, \boldsymbol{\lambda}) = \widehat{X}(\omega, \boldsymbol{\lambda}) + \varepsilon(\omega, \boldsymbol{\lambda}), \quad (\omega, \boldsymbol{\lambda}) \in \mathbb{R}^{d+1}.$$

Here, ε denotes functional spectral Gaussian observation noise, that is, $\varepsilon \in H$, $E[\varepsilon] \equiv 0$, and

$$E[\varepsilon \otimes \varepsilon] = \sigma_\varepsilon^2 \mathbf{I},$$

where $\mathbf{I}$ is the identity operator on $\mathbb{R}^{d+1}$. Process ε is also assumed to be uncorrelated with $\widehat{X}$. By H we denote a separable Hilbert space of spatiotemporal functions, assumed to be included in $L^2(\mathbb{R}^{d+1})$, the space of square integrable function on $\mathbb{R}^{d+1}$. The spectral functional random variable $\widehat{X} \in H$ is defined from the realizations of the spectral process associated with a Gaussian long-range dependence spatiotemporal random field X given by:

$$X(t, \mathbf{z}) \underset{m.s}{=} \int_{\mathbb{R}^{d+1}} r(t - s, \mathbf{z} - \mathbf{y}) Y(s, \mathbf{y}) ds d\mathbf{y}, \tag{1}$$

where

$$r(t, \mathbf{z}) = |t|^{-1+\nu} \prod_{i=1}^{d} |z_i|^{-1+\beta_i}, \tag{2}$$

with $(\nu, \beta_1, \ldots, \beta_d) \in (0, 1/2)^{d+1}$, $t \in \mathbb{R}$, $\mathbf{z} \in \mathbb{R}^d$. Note that, kernel r displays an anisotropic heavy-tail behavior, inducing a local multi-self-similar behavior of our functional spectral data in a neighborhood of zero-frequency. The input spatiotemporal random field Y of model (1) satisfies the following conditions, needed for the pointwise definition of X (see Leonenko (1999), Adler (1981) and Ruiz-Medina, Ángulo, and Anh (2003)):

Condition 1.

$$|f_Y(\omega, \boldsymbol{\lambda})| \longrightarrow C_1,$$

when $\omega \longrightarrow 0$ and $\boldsymbol{\lambda} = (\lambda_1, \ldots, \lambda_d)$, $\lambda_i \longrightarrow 0$, for $i = 1, \ldots, d$, with C_1 being a positive constant and f_Y the spectral density of the spatiotemporal process Y.

Condition 2.

$$\frac{|f_Y(\omega, \boldsymbol{\lambda})|}{(1 + |(\omega, \boldsymbol{\lambda})|^2)^{-\widetilde{\nu} - \sum_{i=1}^{d} \widetilde{\beta}_i}} \longrightarrow C_2,$$

when $\omega \longrightarrow \infty$ and $\lambda_i \longrightarrow \infty$, for $i = 1, \ldots, d$, and for $\boldsymbol{\lambda} = (\lambda_1, \ldots, \lambda_d)$, where C_2 is a positive constant, $(\widetilde{\nu}, \widetilde{\beta}_1, \ldots, \widetilde{\beta}_d) \in (1/2, 1)^{d+1}$, and f_Y denotes, as before, the spectral density of the *input* spatiotemporal process Y.

Remark 1 *Note that Condition 1 means that the integrability order of the spectral density of the spatiotemporal process X at zero frequency depends only of the behavior of the Fourier transform $\hat{r}(\omega, \boldsymbol{\lambda}) = |\omega|^{-\nu} \prod_{i=1}^{d} |\lambda_i|^{-\beta_i}$ of kernel r, at a neighborhood of zero-frequency. This* *behavior is characterized in terms of the range of the parameter vector $(\nu, \beta_1, \ldots, \beta_d)$ (which determines the integrability of the spectral density f_X at the origin). Specifically, the integrability at zero of f_X holds for $(\nu, \beta_1, \ldots, \beta_d) \in (0, 1/2)^{d+1}$. On the other hand, the asymptotic order at infinity of the spectral density f_X of X depends on the considered ranges for vectors $(\nu, \beta_1, \ldots, \beta_d)$ and $(\tilde{\nu}, \tilde{\beta}_1, \ldots, \tilde{\beta}_d)$. In this case, for $\tilde{\nu} > (1/2) - \nu$ and $\tilde{\beta}_i > (1/2) - \beta_i$, for $i = 1, \ldots, d$, f_X is absolutely integrable at infinity.*

3. Estimation methodology

As commented, from Condition 1, the spectral density f_X of spatiotemporal process X displays the following asymptotic local fractal behavior, when $|\omega| \to 0$, and $|\lambda_i| \to 0$, $i = 1, \ldots, d$, (see, Leonenko (1999)),

$$f_X(\omega, \boldsymbol{\lambda}) \sim C_1 |\omega|^{-2\nu} \Pi_{i=1}^d |\lambda_i|^{-2\beta_i}, \ \nu \in (0, 1/2), \ \beta_i \in (0, 1/2), \ i = 1, \ldots, d. \quad (3)$$

In the implementation of the functional estimation algorithms formulated below, compactly supported orthonormal wavelet bases are considered. The multiresolution analysis of $L^2(\mathbb{R}^{d+1})$ is performed in terms of the tensorial product of $d + 1$ one-dimensional orthonormal wavelet bases (see, Meyer (1992)). The one-dimensional wavelet transforms are defined in terms of a scaling basis $\{\phi_k : k \in \Gamma_0 \subset \mathbb{Z}\}$ of a coarsest scale space V_0, and a sequence of wavelet bases $\{\psi_{j:k} : k \in \Lambda_j \subset \mathbb{Z}, \ j \geq 0\}$ of the detail space sequence $\{W_j, \ j \geq 0\}$. The index sets Γ_0 and Λ_j, for $j \geq 0$, are constituted by the integer values k defining the needed translations for covering, at different resolution levels, the one-dimensional zero-frequency neighborhoods considered at the temporal, and at each one of the main spatial directions, in the estimation algorithms described below. Suitable examples of such bases can be constructed from the tensorial product of Haar and Daubechies systems.

To remove the local variability represented by parameter σ_ε^2, due to the measurement noise, universal wavelet thresholding is applied to the empirical wavelet spectral coefficients. That is, we consider the universal threshold UT defined by

$$UT = \sigma_\varepsilon \sqrt{2 \log n},$$

where n denotes the sample size. Here, we have chosen UT, since the noise ε, resulting from the spectral instrument error, is assumed to be Gaussian distributed (see Donoho and Johnstone (1995)). Alternative thresholding rules must be considered for going beyond the Gaussian assumption in order to preserve signal energy, e.g. Lorenz Thresholding (see Vidakovic (1999)).

Under Conditions 1 and 2, the following asymptotic identities are obtained for the wavelet transform of the square-root, $f_X^{1/2}$, of the spectral density f_X, when $R \to 0$,

$$f_{j:k}^1 = \int_{[-R,R]} f_Y^{1/2}(\omega, \boldsymbol{\lambda^0}) |\omega|^{-\nu} \prod_{i=1}^d |\lambda_i^0|^{-\beta_i} \psi_{j:k}(\omega) d\omega \sim 2^{-j(-\nu+1/2)} C(\psi, \boldsymbol{\lambda^0}), \quad (4)$$

for a fixed spatial frequency value $\boldsymbol{\lambda^0}$ in a neighborhood of the spatial zero frequency, where $[-R, R]$ denotes a one-dimensional interval of length $2R$ containing the point zero, and

$C(\psi, \boldsymbol{\lambda^0})$ is a constant depending on the wavelet basis chosen and on the fixed spatial frequency value $\boldsymbol{\lambda^0}$. Specifically, under condition 1,

$$C(\psi, \boldsymbol{\lambda^0}) \sim C_1 \prod_{i=1}^{d} |\lambda_i^0| \int_{[-R,R]} |s|^{-\nu} \psi(s-k) ds.$$

Here, for $k \in \Lambda_j$ and $j \geq 0$, $f_{j:k}^1$ denotes the one-dimensional temporal wavelet coefficient of the square-root $f_X^{1/2}$ of the spectral density f_X of our random field of interest X, evaluated at $\boldsymbol{\lambda^0}$, with respect to the element $\psi_{j:k}$ of the wavelet basis selected. Similarly, when $R \to 0$, we have

$$
\begin{aligned}
f_{j:k}^{1+i} &= \int_{[-R,R]} f_Y^{1/2}(\omega^0, \lambda_1^0, \ldots, \lambda_i, \ldots, \lambda_d^0) \\
&\quad \times |\omega^0|^{-\nu} |\lambda_1^0|^{-\beta_1} \cdots |\lambda_i|^{-\beta_i} \cdots |\lambda_d^0|^{-\beta_d} \psi_{j:k}(\lambda_i) d\lambda_i \\
&\sim 2^{-j(-\beta_i+1/2)} C(\psi, \omega^0, \ldots, \lambda_{i-1}^0, \lambda_{i+1}^0, \ldots, \lambda_d^0), \quad i = 1, \ldots, d,
\end{aligned}
\tag{5}
$$

where, as before, $C(\psi, \omega^0, \lambda_1 \ldots, \lambda_{i-1}^0, \lambda_{i+1}^0, \ldots, \lambda_d^0)$ is a constant depending on the wavelet basis chosen, and on the fixed frequency values $\omega^0, \ldots, \lambda_{i-1}^0, \lambda_{i+1}^0, \ldots, \lambda_d^0$ in a neighborhood of zero frequency. For instance, under Condition 1,

$$
\begin{aligned}
C(\psi, \omega^0, \ldots, \lambda_{i-1}^0, \lambda_{i+1}^0, \ldots, \lambda_d^0) &\sim C_1 |\omega^0|^{-\nu} |\lambda_1^0|^{-\beta_1} \cdots |\lambda_{i-1}^0|^{-\beta_{i-1}} |\lambda_{i+1}^0|^{-\beta_{i+1}} \cdots |\lambda_d^0|^{-\beta_d} \\
&\quad \times \int_{[-R,R]} |s|^{-\beta_i} \psi(s-k) ds.
\end{aligned}
$$

Here, for $k \in \Lambda_j$, $j \geq 0$, and $i = 1, \ldots, d$, $f_{j:k}^{1+i}$ denotes, as in the temporal case, the one-dimensional spatial wavelet coefficient of the square-root

$$
\begin{aligned}
f_X^{1/2}(\omega^0, \ldots, \lambda_{i-1}^0, \cdot, \lambda_{i+1}^0, \ldots, \lambda_d^0) &= f_Y^{1/2}(\omega^0, \ldots, \lambda_{i-1}^0, \cdot, \lambda_{i+1}^0, \ldots, \lambda_d^0) \\
&\quad \times |\omega^0|^{-\nu} \prod_{j=1, j\neq i}^{d} |\lambda_j^0|^{-\beta_j} |\cdot|^{-\beta_i}
\end{aligned}
\tag{6}
$$

of the spectral density f_X of our random field of interest X, with respect to the element $\psi_{j:k}$ of the wavelet basis selected. Note that, under Conditions 1 and 2, f_Y admits a spectral factorization, that is, there exists $f_Y^{1/2}$ satisfying $f_Y = f_Y^{1/2} \overline{f_Y^{1/2}}$, with $\overline{f_Y^{1/2}}$ denoting the complex conjugate. This fact allows the definition, as in equation (6), of the square-root $f_X^{1/2}$ of the spectral density of process X, in terms of the square-root $f_Y^{1/2}$ of f_Y, as well as in terms of the Fourier transform $\hat{r}$ of kernel r. Thus, equations (4)-(5) can be explicitly computed, and the temporal memory parameter ν and spatial dependence parameters β_i, $i = 1, \ldots, d$, can be estimated from the following equations:

$$\log_2 f_{j:k}^1 = -j(-\nu + 1/2) + \log_2 C(\psi, \boldsymbol{\lambda^0}), \tag{7}$$

$$\log_2 f_{j:k}^{1+i} = -j(-\beta_i + 1/2) + \log_2 C(\psi, \omega^0, \ldots, \lambda_{i-1}^0, \lambda_{i+1}^0, \ldots, \lambda_d^0), \tag{8}$$

for $i = 1, \ldots, d$.

The following functional filtering and parameter estimation algorithms are proposed.

Algorithm 1:

Step 1: Define the zero frequency neighborhood sequences: In time

$$[-R_j, R_j] \times [-\boldsymbol{\lambda}^{0,l}, \boldsymbol{\lambda}^{0,l}]^d, \quad j \in \mathbb{N},$$

for each $l = 1, \ldots, L_0$, and $\{\boldsymbol{\lambda}^{0,l}\}_{l=1}^{L_0}$ being a set of fixed spatial frequency vectors with decreasing positive real components, and $R_j \to 0$, as $j \to \infty$. Define also, in the space, for each $i = 1, \ldots, d$, the sequence

$$[-\omega^{0,l}, \omega^{0,l}] \times \cdots \times [-\lambda_{i-1}^{0,l}, \lambda_{i-1}^{0,l}] \times [-R_j, R_j] \times [-\lambda_{i+1}^{0,l}, \lambda_{i+1}^{0,l}] \times, \cdots \times [-\lambda_d^{0,l}, \lambda_d^{0,l}],$$

for $j \in \mathbb{N}$, and with $\{\omega^{0,l}\}_{l=1}^{L_0}$ $\{\lambda_m^{0,l}\}_{l=1}^{L_0}$, for $m = 1, \ldots, d$, $m \neq i$, being decreasing positive numbers sets in the spatial frequency domain, and, as before, $R_j \to 0$, as $j \to \infty$. These sequences of frequency sets provide the supports where the spectral process $\hat{X}$ is evaluated.

Step 2: For each $n \in \mathbb{N}$, compute the average on $l = 1, \ldots, L_0$, of the temporal sample spectral curves, which represent the evaluation of process $\hat{X}$ for $\omega \in [-R_j, R_j]$, for a fixed $\boldsymbol{\xi}^{0,l} \in [-\boldsymbol{\lambda}^{0,l}, \boldsymbol{\lambda}^{0,l}]^d$, and the average of the spatial sample spectral curves, which represent the evaluation of process $\hat{X}$ for $\lambda_i \in [-R_j, R_j]$, for each $i = 1, \ldots, d$, and considering fixed $\tau^{0,l} \in [-\omega^{0,l}, \omega^{0,l}]$, $\xi_m^{0,l} \in [-\lambda_m^{0,l}, \lambda_m^{0,l}]$, $m = 1, \ldots, d$, $m \neq i$, and $l = 1, \ldots, L_0$.

Step 3: Apply the one-dimensional wavelet transform to each element of the averaged temporal and spatial spectral curve sequence obtained in Step 2.

Step 4: Universal wavelet threshold is considered for removing noise in the wavelet coefficients computed in Step 3.

Step 5: At each zero frequency neighborhood, derive from equations (7) and (8), applying linear regression, estimates $\hat{\nu}$ and $\hat{\beta}_i$, $i = 1, \ldots, d$, of the temporal and spatial long-range dependence parameters.

Step 6: The arithmetic mean of the $d + 1$ parameter estimate sequences derived in the previous step is computed.

Algorithm 2:

Step 1: It is given as in Algorithm 1.

Step 2: For each $j \in \mathbb{N}$, and for each $l = 1, \ldots, L_0$, compute the wavelet transform of the temporal sample spectral curves which represent the evaluation of process $\hat{X}$ for $\omega \in [-R_j, R_j]$, and for fixed $\boldsymbol{\xi}^{0,l} \in [-\boldsymbol{\lambda}^{0,l}, \boldsymbol{\lambda}^{0,l}]^d$, and of the spatial sample spectral curves, which represent the evaluation of process $\hat{X}$ for $\lambda_i \in [-R_j, R_j]$, for each $i = 1, \ldots, d$, and for fixed $\tau^{0,l} \in [-\omega^{0,l}, \omega^{0,l}]$, and $\xi_m^{0,l} \in [-\lambda_m^{0,l}, \lambda_m^{0,l}]$, $m = 1, \ldots, d$, $m \neq i$, and $l = 1, \ldots, L_0$.

Step 3: It coincides with Step 4 of Algorithm 1.

Step 4: It coincides with Step 5 of Algorithm 1.

Step 5: At each element of the zero frequency neighborhood sequences, average the temporal and spatial long-range dependence parameter estimates obtained in Step 4.

Step 6: It is defined as in Algorithm 1.

A third estimation algorithm, Algorithm 3, is defined considering in Algorithm 2 a smoothing version of the wavelet transform, with respect to the translation parameter at each resolution level. Finally, the formulation of Algorithm 1 in terms of a smoothing version, over the translation parameter, of the wavelet transform leads to Algorithm 4. As commented, in all the cases, universal wavelet threshold is considered, for removing the observation noise from the functional spectral data (see, for example, Vidakovic (1999)).

4. Simulations

A simulation study is developed to show the performance of the functional estimation algorithms proposed. Two spatiotemporal Gaussian stationary models, defining processes X_1 and X_2, are considered, having the following spectral densities:

$$f_{X_1}(\omega, \lambda_1, \lambda_2) = \left[\frac{1}{(1 + |\omega|^2)^{\alpha_1}}\right] \left[\frac{1}{(1 + |\lambda_1|^2)^{\alpha_2}}\right] \left[\frac{1}{(1 + |\lambda_2|^2)^{\alpha_3}}\right]$$
$$\times \ |\omega|^{-2\nu}|\lambda_1|^{-2\beta_1}|\lambda_2|^{-2\beta_2}, \tag{9}$$

with $\alpha_i \in (1/2, \infty)$, $i = 1, 2, 3$, and

$$f_{X_2}(\omega, \lambda_1, \lambda_2) = \left[\frac{1}{1 + |\omega|^{2\alpha_1}}\right] \left[\frac{1}{1 + |\lambda_1|^{2\alpha_2}}\right] \left[\frac{1}{1 + |\lambda_2|^{2\alpha_3}}\right]$$
$$\times \ |\omega|^{-2\nu}|\lambda_1|^{-2\beta_1}|\lambda_2|^{-2\beta_2}, \tag{10}$$

with $\alpha_i \in (1/2, \infty)$, $i = 1, 2, 3$, respectively. Note that, for $\alpha_i \in (1/2, 3/2)$, the above Gaussian models also display anisotropic fractality. Functional spectral data are constructed from $256 \times 256 \times 256$ frequency points belonging to the interval $[-127.5 * 10^{-8}, 127.5 * 10^{-8}]$, that is, $(\omega, \lambda_1, \lambda_2) \in [-127.5 * 10^{-8}, 127.5 * 10^{-8}]^3$, with discretization step size 10^{-8}. The simulation study is developed considering two structural parameter scenarios, corresponding to heavy and slight spectral singularity, in the range of strong dependence for the two-above introduced spectral models. In relation to the observation noise, we consider the parameter values $\sigma_\varepsilon = 2 * 10^2$ and $\sigma_\varepsilon = 0.05 * 10^2$, which keep a reasonable signal to noise ratio, according to the truncated spectral density values at a zero-frequency spectral neighborhood. That is, we consider the cases:

Case I: $\nu = 0.375$, $\beta_1 = 0.385$, $\beta_2 = 0.395$, $\alpha_1 = 0.7$, $\alpha_2 = 0.8$, $\alpha_3 = 0.9$,
Case II: $\nu = 0.185$, $\beta_1 = 0.195$, $\beta_2 = 0.205$, $\alpha_1 = 0.7$, $\alpha_2 = 0.8$, $\alpha_3 = 0.9$.

From equations (7) and (8) the following estimates can be derived by performing linear regression in the spectral log-wavelet domain:

$$\hat{\nu} = -\hat{\theta}^{d+1} + \frac{1}{2}, \qquad \hat{\beta}_i = -\hat{\theta}^i + \frac{1}{2}, \quad i = 1, \ldots, d,$$

where for $i = 1, \ldots, d+1$, $\hat{\theta}^i$ is computed by applying linear regression. Functional estimation algorithms 1, 2, 3 and 4 are implemented from the following spectral curve sample sizes

$n = 16, 36, 100, 400, 900, 1600, 2500, 3600, 4900, 6400$, at temporal and spatial directions. Figures 1-10 display the long-range dependence parameter estimate sequences, and the estimation of their standard deviation after applying Algorithms 1-4, as well as alternative estimation methods, the partially-integrated method and the marginal-integrated method, previously proposed in Frías, Ruiz-Medina, Alonso, and Angulo (2006) and Frías *et al.* (2009), respectively. Specifically, Case I in model (9), with $\sigma_{\varepsilon_2} = 2 * 10^2$, is displayed in Figures 1 and 2, where $\hat{\nu}$ (red), $\hat{\beta}_1$ (green) and $\hat{\beta}_2$ (blue) values are showed at the top, and standard deviations at the bottom. Dotted line represents the true parameter values. Continuous line provides the results on estimate value sequences after applying hard threholding, i.e., after filtering the spectral data. Dashed line represents the parameter estimate values without previous thresholding.

Figures 3 and 4 show the results obtained for case II and model (9), with $\sigma_{\varepsilon_2} = 2 * 10^2$. The same colors and lines are established for representing estimate value sequences and standard deviations without and with hard thresholding, as well as true parameter values. Figures 5-8 provide the estimation results for: Case I and model (10), with $\sigma_{\varepsilon_2} = 2 * 10^2$; Case II and model (10), with $\sigma_{\varepsilon_2} = 2 * 10^2$. Finally, 9-10 show Case I and model (10) with $\sigma_{\varepsilon_2} = 0.05 * 10^2$, and Case II and model (9) with $\sigma_{\varepsilon_2} = 0.05 * 10^2$.

The results displayed show a better performance of Algorithms 1 and 2 than Algorithms 3 and 4 for high local order of singularity of the spectral density. Additionally, thresholding improves the parameter estimation results in most of the cases considered with Algorithms 1 and 2. Note that, Algorithm 3 and 4 are not compatible with hard-thresholding. They provide better results when the lowest levels of structural and noise local variability are considered. Thus, Algorithms 1 and 2 outperform Algorithms 3 and 4 in the higher local singular cases, and the discrimination between signal and noisy energies, achieved by applying thresholding, improves the parameter estimation results, for a suitable signal to noise ratio. In comparison with the previous estimation methodologies implemented in Frías *et al.* (2006) and Frías *et al.* (2009), the partially-integrated method and the marginal-integrated method, a better performance is obtained with Algorithms 1 and 2, when Case I is considered, in relation to the accuracy of the estimations. It should be noted that estimators with Algorithms 1 and 2 display similar empirical variability properties to those ones presented by previously designed methods in Frías *et al.* (2006) and Frías *et al.* (2009).

5. Application

In this section, the performance of the estimation algorithms proposed is illustrated with a real-data example. The data set studied consists of mean annual daily ocean surface temperature profiles with data size equal to 250 observations, collected from weather stations in the Hawaii Ocean and the West Coast of the United States (latitude-longitude interval $[21.34, 49.98] \times [-158.361, -120, 90]$). In particular, the year 2000 is analyzed at 256 weather stations, according to the availability of the public oceanographic bio-optical database, *The Worldwide Ocean Optics Database (WOOD)*, from which data are collected. The fast Fourier transform is applied to mean annual daily temperature profiles to implement Steps 1-6 of Algorithms 1-4, in terms of the corresponding temporal and spatial spectral curves. Haar wavelet transform is applied in all the cases. Figure 11 shows the long-range dependence parameter estimate sequences, and the estimation of their standard deviations after applying Algorithms 1-4. The spectral curve sample sizes used at temporal and spatial directions are

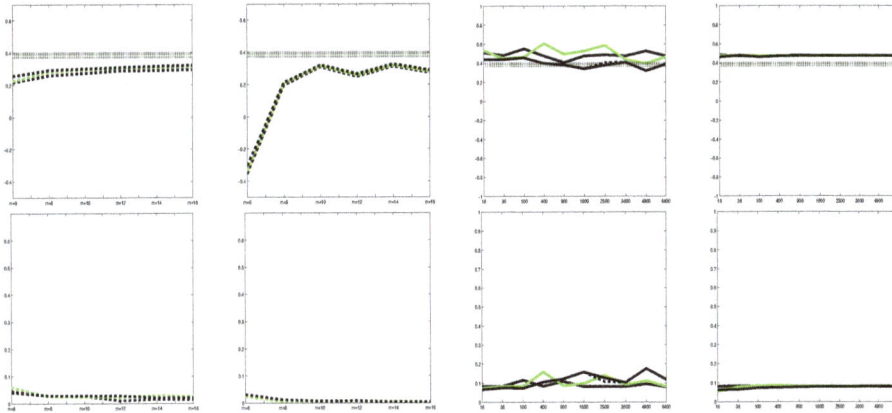

Figure 1: $\hat{\nu}$, $\hat{\beta}_1$, and $\hat{\beta}_2$ values (top) and standard deviations (bottom), partially-integrated method (left), marginal-integrated method (left-medium), algorithm 1 (right-medium), algorithm 2 (right), for case I and for model (9) with $\sigma_{\varepsilon_2} = 2 * 10^2$. The values on horizontal axis represent the spectral curve sample sizes considered.

$\hat{\sigma}(\hat{\nu})$	Algorithm 1	Algorithm 2	Algorithm 3	Algorithm 4
$n = 4$	0.0134	0.0155	0.3253	0.3195
$n = 9$	0.0113	0.0120	0.3406	0.3313
$n = 16$	0.0109	0.0113	0.2951	0.5369
$n = 25$	0.0119	0.0122	0.2876	0.2847
$n = 36$	0.0125	0.0126	0.2556	0.2551
$n = 49$	0.0123	0.0125	0.2431	0.2462
$n = 64$	0.0110	0.0111	0.2127	0.2198
$n = 81$	0.0200	0.0201	0.2014	0.2304
$n = 100$	0.0182	0.0182	0.1886	0.2125
$n = 121$	0.0167	0.0168	0.1783	0.1986

Table 1: Standard deviation of $\hat{\nu}$ values.

$n = 4, 9, 16, 25, 36, 49, 64, 81, 100, 121$. Standard deviations of $\hat{\nu}$, $\hat{\beta}_1$ and $\hat{\beta}_2$ are also displayed in Tables 1-3. As showed in the simulation study, Algorithms 1 and 2 are more efficient than Algorithms 3 and 4. Note that the spatial strong dependence is induced by the high concentration level of weather stations. Since this concentration is similar in terms of longitude and latitude magnitudes, the two spatial long-range dependence parameters are very close. This fact is reflected in the results displayed in Figure 11.

For comparative purposes, considering the same functional data sets, the previous spectral-based estimation methodologies proposed in Frías et al. (2006) and Frías et al. (2009) are also applied for estimation of the temporal and spatial long-range dependence parameters in the spectral domain (see Figure 12). As in the simulation study, it can be appreciated that similar empirical variability properties are displayed by the estimates computed with the previous spectral methods in Frías et al. (2006) and Frías et al. (2009), and with Algorithms 1 and 2 in the spectral- wavelet domain. However, although similar estimates are obtained, with both methodologies, for the spatial long-range dependence parameters, bigger differences are

$\hat{\sigma}(\hat{\beta}_1)$	Algorithm 1	Algorithm 2	Algorithm 3	Algorithm 4
$n = 4$	0.0399	0.0299	0.0020	0.0019
$n = 9$	0.0359	0.0265	0.0033	0.0032
$n = 16$	0.0326	0.0226	0.0029	0.0029
$n = 25$	0.0268	0.0198	0.0032	0.0032
$n = 36$	0.0224	0.0174	0.0068	0.0067
$n = 49$	0.0207	0.0165	0.0095	0.0094
$n = 64$	0.0175	0.0143	0.0123	0.0123
$n = 81$	0.0164	0.0133	0.0113	0.0112
$n = 100$	0.0151	0.0123	0.0103	0.0103
$n = 121$	0.0142	0.0115	0.0095	0.0095

Table 2: Standard deviation of $\hat{\beta}_1$ values.

$\hat{\sigma}(\hat{\beta}_2)$	Algorithm 1	Algorithm 2	Algorithm 3	Algorithm 4
$n = 4$	0.0087	0.0089	0.0050	0.0054
$n = 9$	0.0080	0.0077	0.0042	0.0048
$n = 16$	0.0066	0.0066	0.0038	0.0044
$n = 25$	0.0059	0.0060	0.0045	0.0045
$n = 36$	0.0054	0.0055	0.0048	0.0048
$n = 49$	0.0053	0.0052	0.0050	0.0050
$n = 64$	0.0048	0.0047	0.0047	0.0046
$n = 81$	0.0045	0.0044	0.0046	0.0046
$n = 100$	0.0041	0.0041	0.0046	0.0048
$n = 121$	0.0038	0.0038	0.0044	0.0045

Table 3: Standard deviation of $\hat{\beta}_2$ values.

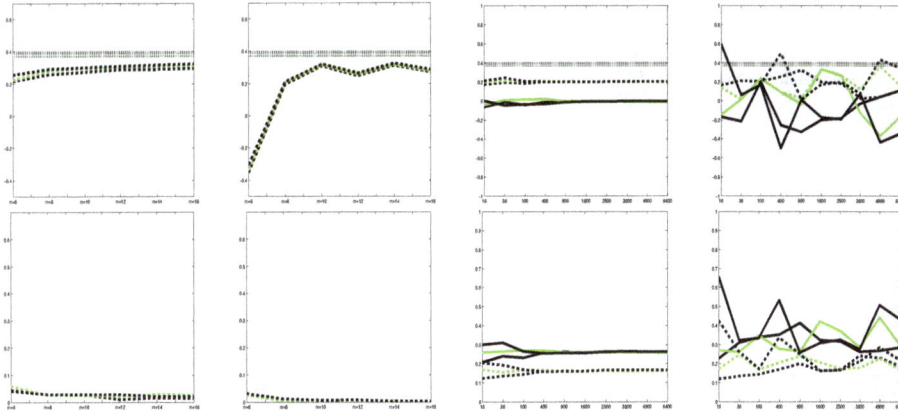

Figure 2: $\hat{\nu}$, $\hat{\beta}_1$, and $\hat{\beta}_2$ values (top) and standard deviations (bottom), partially-integrated method (left), marginal-integrated method (left-medium), algorithm 3 (right-medium), algorithm 4 (right), for case I and for model (9) with $\sigma_{\varepsilon_2} = 2 * 10^2$. The values on horizontal axis represent the spectral curve sample sizes considered.

appreciated in the temporal long-range dependence parameter estimates. After computing the corresponding least-squares plug-in estimators of the values of the original process, it is observed a better performance of the plug-in predictor based on the parameter estimation in the spectral-wavelet domain. Note that the order of magnitude of the absolute error is less than 0.1 with the proposed methodology, while with the previous spectral methods in Frías et al. (2006) and Frías et al. (2009) is always larger than 0.1. Therefore, we conclude, as in the simulation study, that for strong-dependence parameter values close to Case I, a better performance is obtained with the proposed methodology in the spectral-wavelet domain, in terms of Algorithms 1 and 2.

6. Final comments

Long-range dependence is a key feature in the analysis of complex systems which can be equivalently studied, thanks to Tauberian-type theorems (see, for example, Leonenko (1999)), in terms of the local singularity level, in a neighborhood of the zero frequency, of the spectral density. This fact motivates the parameter estimation methodology proposed in this paper, in terms of compactly supported wavelet functions. The estimation algorithms proposed combine thresholding in the wavelet domain with linear regression from the log-thresholded-wavelet transform of the spectral data. This double filtering is advisable in the cases showed in the simulation study, thus, for processing spectral data displaying high local singularity, due to the heavy tail behavior of the covariance function, and to the local variability of the observation noise. As main conclusion conducted from the paper, we then have that Algorithms 1 and 2 are suitable for processing high local structural spectral variability, corresponding to the range $(0.3, 0.5)$, after applying hard-thresholding in the presence of measurement noise.

On the other hand, the consistency of the wavelet-spectral based estimators formulated for the approximation of the long-range dependence parameters follows from the consistency of the wavelet periodogram, computed from the functional spectral data (see Frías and Ruiz-Medina (2012)). Thus, the efficiency and consistency of these estimators, with a suitable

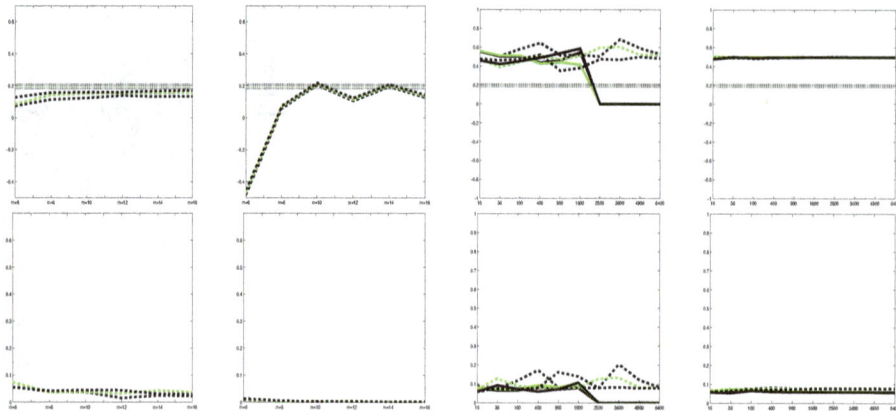

Figure 3: $\hat{\nu}$, $\hat{\beta}_1$, and $\hat{\beta}_2$ values (top) and standard deviations (bottom), partially-integrated method (left), marginal-integrated method (left-medium) algorithm 1 (right-medium), algorithm 2 (right), for case II and for model (9) with $\sigma_{\varepsilon_2} = 2 * 10^2$. The values on horizontal axis represent the spectral curve sample sizes considered.

scaling, ensure a good performance when the functional sample size increases.

Acknowledgments. This work has been supported in part by projects MTM2009-13393 of the DGI, MEC, and P09-FQM-5052 of the Andalousian CICE, Spain.

References

Adler RJ (1981). *The Geometry of Random Fields.* Wiley, London.

Akkaya AD, Yücemen MS (2002). "Stochastic modeling of earthquake occurrences and estimation of seismic hazard: a random field approach." *Probabilistic Engineering Mechanics*, pp. 1–13.

Anh VV, Lam KC, Leung Y, Tieng Q (2000). "Multifractal analysis of Hong Kong air quality data." *Environmetrics*, pp. 139–149.

Baladandayuthapani V, Mallick B, Hong M, Lupton J, Turner N, Caroll R (2008). "Bayesian hierarchical spatially correlated functional data analysis with application to colon carcinoginesis." *Biometrics*, pp. 64–73.

Bardet JM, Lang G, Oppenheim G, Philippe A, Stove S, Taqqu MS (2003). "Semi-parametric estimation of the long-range dependence parameter." *In: Doukhan P, Oppenheim G, Taqqu M (Eds) Theory and Applications of Long-Range Dependence.*

Basse M, Diop A, Dabo-Niang S (2008). *Mean squares properties of a class of kernel density estimates for spatial functional random variables.* Technical Report, Université Gaston Berger de Saint-Louis, Senegal.

Bell JH, BA BB, Martini (2010). "Imaging spectroscopy of jarosite cement in the Jurassic Navajo Sandstone." *Remote Sensing of Environment*, pp. 2259–2270.

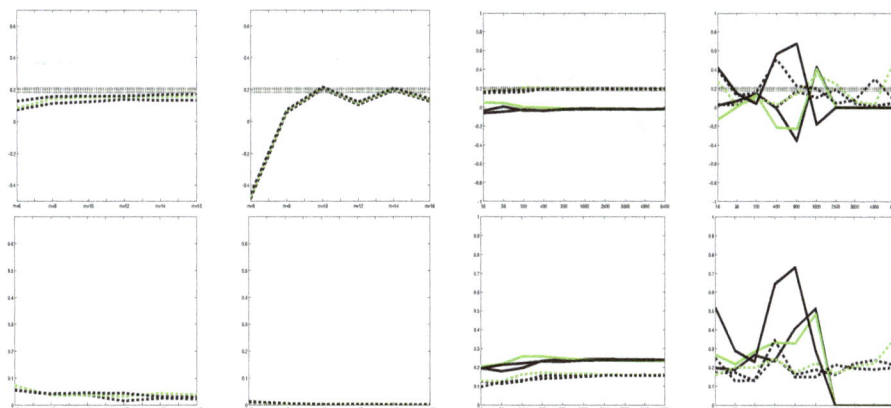

Figure 4: $\hat{\nu}$, $\hat{\beta}_1$, and $\hat{\beta}_2$ values (top) and standard deviations (bottom), partially-integrated method (left), marginal-integrated method (left-medium) algorithm 3 (right-medium), algorithm 4 (right), for case II and for model (9) with $\sigma_{\varepsilon_2} = 2 * 10^2$. The values on horizontal axis represent the spectral curve sample sizes considered.

Benhenni K, Hedli-Griche S, Rachdia M, Vieu P (2008). "Consistency of the regression estimator with functional data under long memory conditions." *Statistics and Probability Letters*, pp. 1043–1049.

Byambakhuu I, Sugita M, Matsushima D (2010). "Spectral unmixing model to assess land cover fractions in Mongolian steppe regions." *Remote Sensing of Environment*, pp. 2361–2372.

Clark RN, Roush TL (1984). "Reflectance Spectroscopy: Quantitative analysis techniques for remote sensing applications." *Journal of Geophysical Research*, pp. 6329–6340.

Delicado P, Giraldo R, Comas C, Mateu J (2009). "Statistics for spatial functional data: some recent contributions." *Environmetrics*, pp. 224–239.

Donoho DL, Johnstone IM (1995). "Adapting to unknown smoothness via wavelet shrinkage." *Journal of the American Statistical Association*, pp. 1200–1224.

Elvidge CD, Chen Z, Groeneveld DP (1993). "Detection of trace quantities of green vegetation in 1990 AVIRIS data." *Remote Sensing of Environment*, pp. 271–279.

Ferraty F, Goia A, Vieu P (2002). "Functional nonparametric model for time series: a fractal approach for dimension reduction." *Test*, pp. 317–344.

Frías MP, Ruiz-Medina MD (2012). "Wavelet-Based Estimation of Anisotropic Spatiotemporal Long-Range Dependence." *Methodology and Computing in Applied Probability (in progress, submitted the revised version)*.

Frías MP, Ruiz-Medina MD, Alonso FJ, Angulo JM (2006). "Semiparametric Estimation of Spatiotemporal Anisotropic Long-Range Dependence." In *Proceedings in Computational Statistics (Contributed Paper)*.

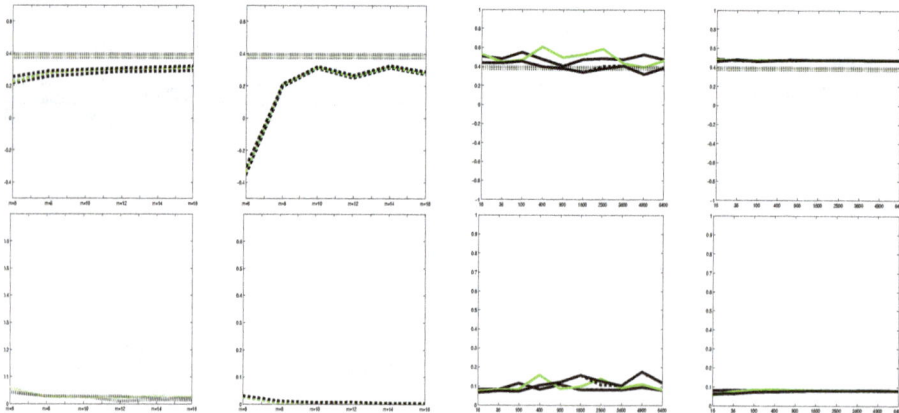

Figure 5: $\hat{\nu}$, $\hat{\beta}_1$, and $\hat{\beta}_2$ values (top) and standard deviations (bottom), partially-integrated method (left), marginal-integrated method (left-medium), algorithm 1 (right-medium), algorithm 2 (right), for case I and for model (10) with $\sigma_{\varepsilon_2} = 2 * 10^2$. The values on horizontal axis represent the spectral curve sample sizes considered.

Frías MP, Ruiz-Medina MD, Alonso FJ, Angulo JM (2006a). "Spatiotemporal Generation of Long-Range Dependence Models and Estimation." *Environmetrics*, **17**, 139–146.

Frías MP, Ruiz-Medina MD, Alonso FJ, Angulo JM (2008). "Parameter Estimation of Self-Similar Spatial Covariogram Models." *Commputation Statatistics - Theory and Methods*, **37**, 1011–1023.

Frías MP, Ruiz-Medina MD, Alonso FJ, Angulo JM (2009). "Spectral-Marginal-Based Estimation of Spatiotemporal Long-Range Dependence." *Commputation Statatistics - Theory and Methods*, **38**, 103–114.

Gallo K, McNab AL, Karl TR, Brown JF, Hood JJ, Tarpley JD (1993). "The use of a vegetation index for assessment of the urban heat island effect." *International Journal of Remote Sensing*, **14,11**, 2223–2230.

Goetz AFH, Vane G, Solomon JE, Rock BN (1985). "Imaging spectrometry for earth remote sensing." *Journal of Time Series Analysis*, **228**, 1147–1153.

Kelbert M, Leonenko N, Ruiz-Medina MD (2005). "Fractional Random Fields Associated with Stochastic Fractional Heat Equation." *Advances in Applied Probability*, **37**, 108–133.

Leonenko N (1999). *Limit Theorems for Random Fields with Singular Spectrum. Mathematics and its Applications*. Kluwer Academic Publishers, Dordrecht, Boston, London.

Marguerit C, Schertzed D, Schmitt F, Lovenjoy S (1998). "Copepod diffusion within multifractal phytoplankton fields." *Journal of Marine Systems*, **16**, 69–83.

Meyer Y (1992). *Wavelet and Operators*. Cambridge University Press, Cambridge.

Price JC (1990). "Using spatial context in satellite data to infer regional scale evapotranspiration." *I.E.E.E. Transactions on Geoscience and Remote Sensing*, **28**, 940–948.

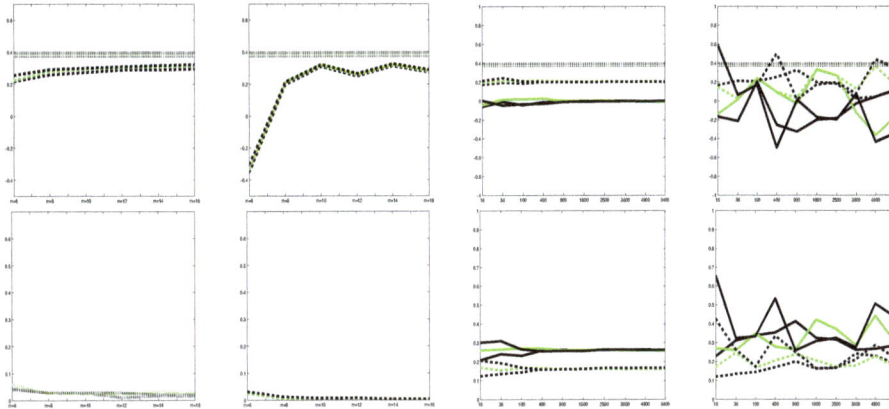

Figure 6: $\hat{\nu}$, $\hat{\beta}_1$, and $\hat{\beta}_2$ values (top) and standard deviations (bottom), partially-integrated method (left), marginal-integrated method (left-medium), algorithm 3 (right-medium), algorithm 4 (right), for case I and for model (10) with $\sigma_{\varepsilon_2} = 2 * 10^2$. The values on horizontal axis represent the spectral curve sample sizes considered.

Ruiz-Medina MD (2011). "Spatial autoregressive and moving average Hilbertian processes." *Journal of Multivariate Analysis*, **102**, 292–305.

Ruiz-Medina MD, Ángulo JM, Anh V (2003). "Fractional generalized random fields on bounded domains." *Stochastics Analysis and Applications*, **21**, 465–492.

Vidakovic B (1999). *Statistical Modeling by Wavelets*. Wiley Series in Probability and Statistics.

Affiliation:

María Dolores Ruiz-Medina
Faculty of Sciences
University of Granada
Campus Fuente Nueva s/n
18071 Granada , Spain
E-mail: mruiz@ugr.es

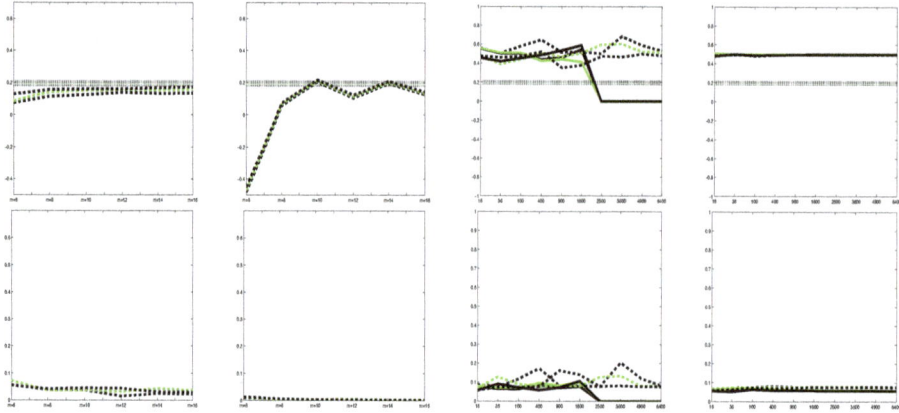

Figure 7: $\hat{\nu}$, $\hat{\beta}_1$, and $\hat{\beta}_2$ values (top) and standard deviations (bottom) partially-integrated method (left), marginal-integrated method (left-medium), algorithm 1 (right-medium), algorithm 2 (right), for case II and for model (10) with $\sigma_{\varepsilon_2} = 2 * 10^2$. The values on horizontal axis represent the spectral curve sample sizes considered.

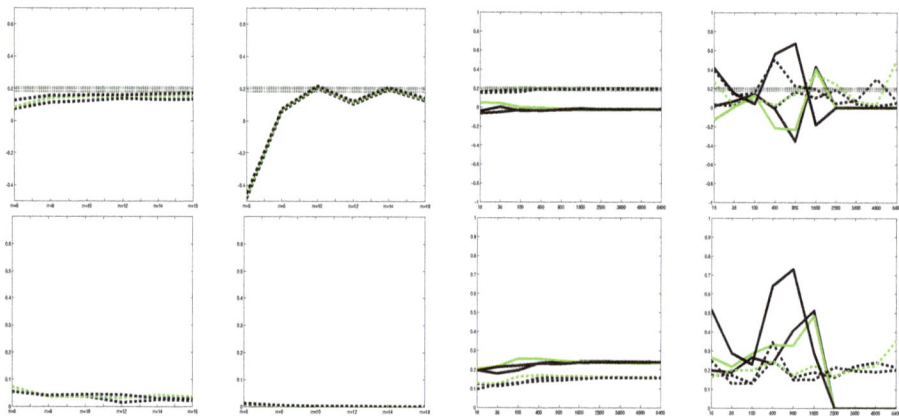

Figure 8: $\hat{\nu}$, $\hat{\beta}_1$, and $\hat{\beta}_2$ values (top) and standard deviations (bottom) partially-integrated method (left), marginal-integrated method (left-medium), algorithm 3 (right-medium), algorithm 4 (right), for case II and for model (10) with $\sigma_{\varepsilon_2} = 2 * 10^2$. The values on horizontal axis represent the spectral curve sample sizes considered.

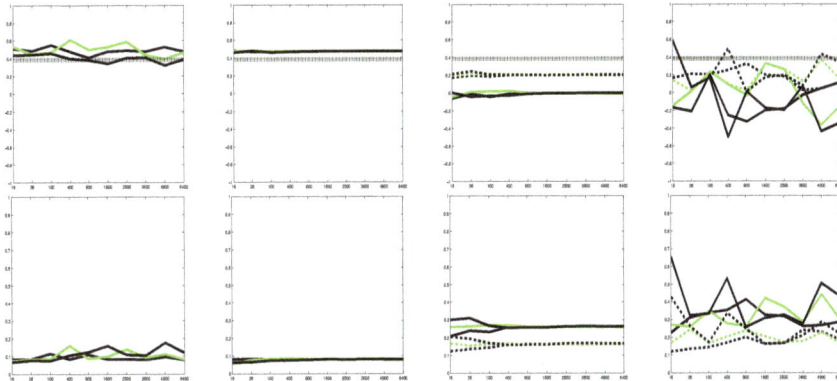

Figure 9: $\hat{\nu}$, $\hat{\beta}_1$, and $\hat{\beta}_2$ values (top) and standard deviations (bottom), algorithm 1 (left), algorithm 2 (left-medium,), algorithm 3 (right-medium), algorithm 4 (right), for case I and for model (10) with $\sigma_{\varepsilon_2} = 0.05 * 10^2$. The values on horizontal axis represent the spectral curve sample sizes considered.

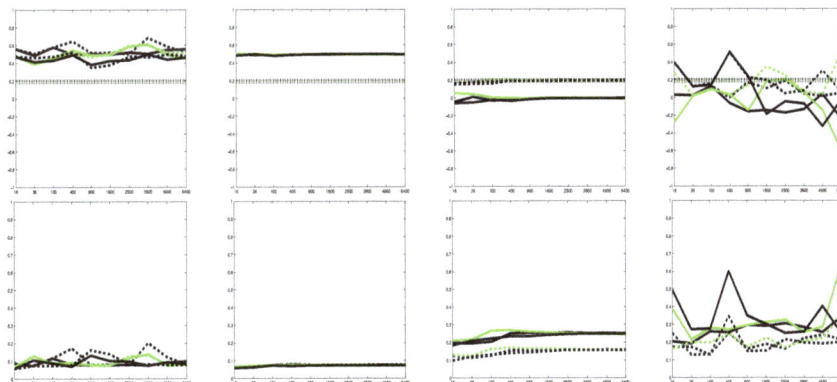

Figure 10: $\hat{\nu}$, $\hat{\beta}_1$, and $\hat{\beta}_2$ values (top) and standard deviations (bottom) algorithm 1 (left), algorithm 2 (left-medium,), algorithm 3 (right-medium), algorithm 4 (right), for case II and for model (9) with $\sigma_{\varepsilon_2} = 0.05 * 10^2$. The values on horizontal axis represent the spectral curve sample sizes considered.

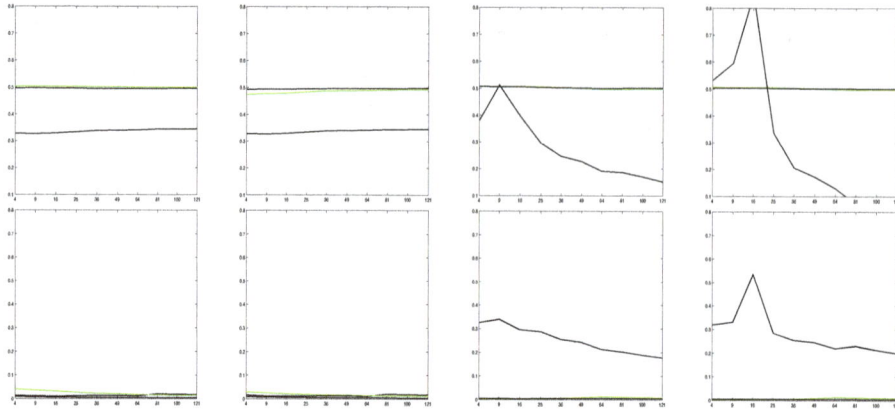

Figure 11: $\hat{\nu}$ (red), $\hat{\beta}_1$, (green) and $\hat{\beta}_2$ (blue) values (top) and standard deviations (bottom), algorithm 1 (left), algorithm 2 (left-medium), algorithm 3 (right-medium), algorithm 4 (right). The values on horizontal axis represent the spectral curve sample sizes considered.

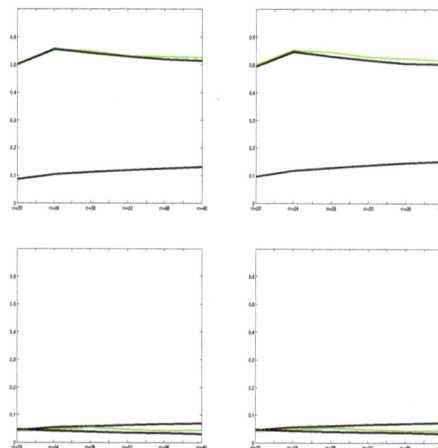

Figure 12: $\hat{\nu}$ (red), $\hat{\beta}_1$ (green) and $\hat{\beta}_2$ (blue) values (top) and standard deviations method (left), marginal-integrated method (right).

14

Characterization Theorems Based on Conditional Quantiles with Applications

Ratan Dasgupta
Indian Statistical Institute

Abstract

We prove a characterization of Pareto variable based on quantiles when conditional distribution above a threshold is considered. A similar characterization for exponential distribution is also obtained. The results are extended to discretized random variables. For some well known distributions, effect of conditioning the variable crossing a threshold on quantiles is investigated. The results are further extended to bivariate exponential and bivariate Pareto type models which are relevant to explain lifestyle data. Applications of the results are made in estimation of conditional quantiles in environmental data. Pareto model for excess of flood-peaks of a river seems to be satisfactory with high threshold values. Applications are also made on Yam-yield, wind speed data of high energy due to extratropical cyclones in coastal regions and worldwide earth-tremor data.

Keywords: Pareto model, quantiles, Cauchy functional equation, lifestyle data, elephant foot yam, extratropical cyclones, peak gust wind.

1. Introduction

Pareto distribution and its variants have wide applications in modeling different branches of science, especially economics. Apart from explaining the distribution of wealth or income, this distribution may explain observed phenomena in sociology, anthropology, hydrology, meteorology, actuarial science, occupational health and safety etc.
See e.g., Cebrian, Denuit, and Lambert (2003), Dasgupta (2011), Jenkinson (1955),
Klass, Biham, Levy, Malcai, and Soloman (2006), Krishnaji (1970),
de Oliveira, Ebecken, de Oliveira, and Gilleland (2011),
Morrow-Tlucak, Emhart, Sokol, Martier, and Ager (1989), Van Montfort and Witter (1985).

In this paper we prove a characterization of Pareto distribution based on quantiles when the variable exceeds a threshold. Comparisons are made between the unrestricted quantiles of original variable, and restricted quantiles for the variable above a threshold. The relationship of constant ratio of unrestricted and restricted quantiles of variable beyond a threshold is seen to characterize the Pareto distribution. The proof involves solving functional equations, like Cauchy functional equation, over a restricted zone. A similar characterization based on constant shift of restricted quantiles from unrestricted quantiles is proved for exponential random variable. The results are generalized for discretized version of the random variables. Effect of conditioning the variable above a threshold on quantiles is discussed for some well known distributions including normal distribution. The case when underreporting of a variable is of exponential order that seems realistic in some specific situations is also studied. We further study the conditional quantiles to obtain relevant characterizations for bivariate exponential and bivariate Pareto type models those are useful in explaining lifestyle data. Applications of the results are made in different contexts including environmental data.

In section 2 we prove the results for Pareto distributions and exponential distributions. Similar characterizations for discrete random variables are proved in section 3. In section 4 we obtain results for bivariate exponential and bivariate Pareto type distributions. Section 5 discusses applications of the results in agricultural data of yam yield, environmental data on flood-peak, earth-tremor and peak gust (PGU) wind velocity. The value of median PGU exceeding the recorded maximum is estimated based on unrestricted median and recorded maximum peak gust, thus providing a glimpse of the scenario beyond observed range.

2. Characterization of Pareto and exponential distribution

We first prove the following.

Theorem 1. Let X be a random variable with support $(a, \infty), a > 0$, and distribution function F. Denote $c = c(p)$ to be the unrestricted p-th quantile of $X(> a)$, and consider p in a (small) dense neighborhood A_0 of origin (e.g., $p \in A_0 = (0, \epsilon) \cap Q, \epsilon > 0$, small and Q is the set of rational numbers). Then the p-th quantile of the distribution, $p \in A_0$, under the restriction $X > x_0(> a)$ is cx_0/a iff F is a Pareto distribution function.

Proof. Consider the distribution function of standardised Pareto variable with $a = 1$.

$$F(x) = 1 - x^{-\alpha}, \ x > 1, \ \alpha > 0 \qquad \qquad ...(2.1)$$

The median of the distribution is at $2^{1/\alpha}$. Denote $\overline{F} = 1 - F$, $g(x) = \log \overline{F}(x) = -\alpha \log x \downarrow -\infty$, $x \uparrow \infty$. The c.d.f. of the variable, given that $x > x_o(> 1)$, then turns out to be $F(x)/\overline{F}(x_0)$, and one may write $P(X > x | X > x_0) = \frac{\overline{F}(x)}{\overline{F}(x_0)} = (\frac{x}{x_o})^{-\alpha}$. Equating this to 0.5 we obtain the new median of the random variable crossing the threshold x_0 as cx_0, where $c = 2^{1/\alpha}$ is the median of the random variable $X(> 1)$.
This property specifies the form of the distribution at the points $c, c^2, \cdots, c^m, \cdots$ as explained below.
For a general distribution function $F = F(x)$ of the random variable $X > 1$, denote $g(x) = \log \overline{F}(x) = \log(1 - F(x))$. Suppose that the new median of the random variable X under the

restriction $x > x_0$ is at cx_0, where c is independent of x_0. Indeed c is the median of original unrestricted random variable as seen by taking $x_0 \downarrow 1$. Next, write

$$e^{g(cx_0)-g(x_0)} = \frac{\overline{F}(cx_0)}{\overline{F}(x_0)} = 0.5 \qquad \qquad ...(2.2)$$

This provides,

$$g(cx_0) - g(x_0) = -k \qquad \qquad ...(2.3)$$

where, $k = \log 2$.

Thus $g(c^2) = g(c) - k = -2k$, $g(c^3) = -3k, \cdots, g(c^m) = -mk$. This implies the type of the distribution function is Pareto, $g(x) = \log \overline{F}(x) = -\alpha \log x$; where $\alpha = k/(\log c)$ at the points $x = c, c^2, \cdots, c^m, \cdots$

Note that a similar relation holds for the third quartile of the Pareto distribution (2.1) with $c = 4^{1/\alpha}$, $k = \log 4$.

Thus equations (2.2)-(2.3) for third quartile of a general F imply Pareto distribution for some other points $x = c, c^2, \cdots, c^m, \cdots$ with a different choice of c.

Now assume that the above property of constant multiple factor of restricted and unrestricted quantiles holds for a dense set of quantiles corresponding to $p \in (0,1)$, p rational. The form of difference equation then reduces to

$$g(cx_0) - g(x_0) = \log(1-p) = -k \qquad \qquad ...(2.4)$$

$k = -\log(1-p) > 0$, $c = (1-p)^{-1/\alpha}$; $p \in Q \cap (0,1)$.

This specifies the distribution function F to be Pareto in a dense set $x = c, c^2, \cdots, c^m, \cdots$, of $(1, \infty)$. For an arbitrary real number $z > 1$, there exist integer m and $c = (1-p)^{-1/\alpha}$; $p \in Q \cap (0,1)$ such that c^m is arbitrary close to the number z, where Q is the set of all rational numbers. Next from right continuity of distribution function, the form of F is Pareto at z, where $z > 1$ is arbitrary.

Finally, a dense choice of p in a small neighborhood of origin, e.g., $p \in A_0 = (0, \epsilon) \cap Q, \epsilon > 0$, small suffices for the Theorem to hold; as the resultant sequence $\{c^m : m = 1, 2, 3, \cdots\}$ still spans a dense support of the variable.

For the general case let the minimum possible value of X be $a > 0$. The Pareto distribution function F with minimum value a is then

$$F(x) = 1 - (x/a)^{-\alpha}, \; x > a(> 0), \; \alpha > 0 \qquad \qquad ...(2.5)$$

One may then consider the transformed random variable $X/a(> 1)$. Proceeding as before the characterization of Theorem 1 holds.

Next we state a similar result for exponential variable.

Theorem 2. Let X be a random variable with support $(a, \infty), a \geq 0$, and distribution function F. Denote $c = c(p)$ to be the unrestricted p-th quantile of $X(> a)$, and consider p in a (small) dense neighborhood A_0 of origin (e.g., $p \in A_0 = (0, \epsilon) \cap Q, \epsilon > 0$, small and Q is the set of rational numbers). Then the p-th quantile of the distribution, $p \in A_0$, under the restriction $X > x_0(> a)$ is $c + x_0 - a$ iff F is an exponential distribution function.

Proof. For exponential random variable Y with distribution function

$$G(y) = 1 - e^{-\lambda y}, y > 0 \qquad \qquad ...(2.6)$$

it is easy to see that the p-th quantile of the distribution under the restriction $Y > y_0(> 0)$ is merely a shift of the unrestricted quantile by y_0.

$$P(Y > y|Y > y_0) = e^{-\lambda(y-y_0)} = 1 - p \Rightarrow y = y_0 - \frac{1}{\lambda}\log(1-p) = y_0 + \xi_Y(p) \qquad ...(2.7)$$

where $\xi_Y(p) = -\frac{1}{\lambda}\log(1-p)$ is the p-th quantile of $Y > 0$.

This property characterizes the exponential distribution.
To see this for a general random variable Y with distribution function G and $y > y_0(> 0)$, assume that $(y_0 + c)$ to be the new p-th quantile; shifted from unrestricted p-th quantile c by y_0. Then write in a similar fashion as in (2.4),

$$e^{g(y_0+c)-g(y_0)} = \frac{1 - G(y_0 + c)}{1 - G(y_0)} = P(Y > y_0 + c|Y > y_0) = 1 - p = e^{\log(1-p)} = e^{-k} \qquad ...(2.8)$$

leading to the equation

$$g(y_0 + c) - g(y_0) = -k \qquad \qquad ...(2.9)$$

where $g(x) = \log \overline{G}(x), \overline{G} = 1 - G$.
One may solve (2.9) in a similar fashion as in (2.4), with the resultant solution of the form $g(x) = -\lambda x$, leading to the exponential distribution. To see this write $g(mc) = g((m-1)c) - k = \cdots = -mk$, and g is seen to be linear on the points $c, 2c, 3c, \cdots, mc, \cdots$ thus implying exponential distribution at those points. Theorem 2 is then immediate following similar steps of proof as in Theorem 1.
A dense choice of p in a small neighborhood of origin, $A_0 = (0, \epsilon) \cap Q, \epsilon > 0$, small suffices for the Theorem to hold; the resultant sequence $\{mc : m = 1, 2, 3, \cdots\}$; $c = c(p), p \in A_0$ still spans a dense support of the variable.

Equation (2.9) is seen to be a variant of equation (2.4). Write $f(x) = g(e^x)$, then from (2.4), $g(e^{\log c + \log x_0}) - g(e^{\log x_0}) = -k$. That is $f(x) = g(e^x)$, is of the form (2.9) in log scale as $f(\log x_0 + \log c) - f(\log x_0) = -k$.

Pareto and exponential distributions are inter related as follows. If X is Pareto-distributed with minimum a and index δ, then $Y = \log(X/a)$ is exponentially distributed with intensity δ. Equivalently, if Y is exponentially distributed with intensity δ, then ae^Y is Pareto-distributed with minimum a and index δ. This relationship is reflected in the similarity of equations (2.4) and (2.9).

Remark 1. Equations (2.4) and (2.9) are related to Cauchy functional equation. The constants in the r.h.s. of these two equations are $-k = \log(1-p) = g(c)$. Thus these two equations can be rewritten in the form $g(cx_0) = g(c) + g(x_0)$ and $g(y_0 + c) = g(y_0) + g(c)$, respectively. As already mentioned (2.4) and (2.9) are reformulations of each other. Variation of x_0 is due to shift of threshold, the other coordinate c varies as $p \in (0, \epsilon) \cap Q$ varies.

Apart from some pathological examples, the solutions of Cauchy functional equation $g(x+y) = g(x) + g(y)$ over R or R^+ is of the form $g(x) = \lambda x$.

In the present case $g(x) = \log(1 - F(x))$ is a monotone function on R^+.

Remark 2. The change in the value of quantile is a result of conditioning the random variable towards the tail of the distribution. When the tail is moderately decaying like exponential then shift in quantile equals shift in the threshold of the random variable. However, for a thick tailed distribution like Pareto with polynomial decay, shift of quantile is high towards tail; it is a constant (> 1) multiple of original quantile. In this context it is worthwhile to examine some other distributions and the status of normal distribution in the scenario. For exponential distribution (2.6) note that

$$\frac{\overline{G}(y)}{\overline{G}(y_0)} = \frac{e^{-\lambda y}}{e^{-\lambda y_0}} = \left(\frac{x}{x_0}\right)^{-\lambda} \qquad ...(2.10)$$

writing $e^y = x$. Equating the r.h.s to $(1 - p)$ the value of restricted quantiles are obtained. The above also shows the interrelation of exponential distribution with Pareto distribution and the corresponding shifts of quantiles when crossing of threshold $x_0 = e^{y_0}$ is considered for the transformed variable $X = e^Y$. Tail probability of Pareto variable X decays at slower rate $O(x^{-\lambda})$ compared to exponential decay $O(e^{-\lambda y})$ for the exponential variable Y. As a result restricted quantile of Pareto is wide apart from unrestricted quantile, compared to that for exponential variable.

Below we check the effect of crossing a (large) threshold on the quantiles of some other distributions.

1. Normal distribution. For a standardized normal variable Z

$$P(Z > z)/P(Z > z_0) = \Phi(-z)/\Phi(-z_0) \sim (z/z_0)^{-1} e^{-(z^2 - z_0^2)/2} \qquad ...(2.11)$$

where $z > z_0 (> 0)$, and z_0 is large. Thus an approximate value of the restricted p-th quantile for standardized normal distribution having crossed a high threshold $z_0(> 0)$ is given by the following

$$z \approx [-2\log(1 - p) + z_0^2]^{1/2} \sim z_0[1 - \frac{1}{z_0^2}\log(1 - p)] \qquad ...(2.12)$$

From r.h.s. of (2.12) it is seen that the restricted p-th quantile tends to z_0 for large value of the threshold z_0.

For exponential distribution with relatively thick tail the difference between the restricted and unrestricted quantiles, as we have seen earlier, is the amount of shift in threshold value, i.e., the difference between the restricted quantile and the threshold value is a constant, viz., the unrestricted p-th quantile; irrespective of the value of threshold. However, for normal distribution with relatively fast decaying tail the difference $z - z_0 \approx -\frac{1}{z_0}\log(1 - p) \to 0$, as $z_0 \to \infty$.

2. *Weibull distribution.* The standard Weibull distribution have cumulative distribution function

$$H(v) = 1 - e^{-v^k}, v \geq 0, \ k > 0 \qquad \qquad \text{...(2.13)}$$

Solving the equation

$$\frac{\overline{H}(v)}{\overline{H}(v_0)} = \frac{e^{-v^k}}{e^{-v_0^k}} = (1-p) \qquad \qquad \text{...(2.14)}$$

for $v > v_0(> 0)$, one gets the restricted p-th quantile for Weibull distribution having crossed the threshold v_0 as, $v = [v_0^k - \log(1-p)]^{1/k}$.

For $k > 1$, $v \approx v_0[1 - \frac{1}{kv_0^k}\log(1-p)]$, and conclusion similar to normal distribution holds in this case.

The distribution (2.13) for $0 < k < 1$ has a lower order decay of tail probability than exponential distribution ($k = 1$), and the above analysis indicates that the difference between restricted and unrestricted qualtiles is more than the shift in threshold, whereas with $k > 1$ tail probability decays faster than exponential distribution, and the difference between restricted and unrestricted quantiles shrinks towards zero as the value of the threshold increases towards infinity.

3. *Exponential underreporting and Pareto model.* Underreporting to a high level of a variable may change the pattern of distribution of the variable of interest. The phenomenon of underreporting is present in many occasions like traffic injuries, HIV infection, credit card debt etc. In some cases it may be to the tune of 20 fold, e.g., gross underreported alcohol use in pregnancy, see de Oliveira *et al.* (2011). A model of exponential underreporting may then be more appropriate compared to underreporting to a multiplicative factor. Although in most of the present studies we consider income /wealth underreporting up to a multiplicative factor, it would be interesting to see how the model and relevant analysis change from traditional Pareto model (2.1), if we take into account the possibility of exponential underreporting for some specific cases as reported in de Oliveira *et al.* (2011). To this end consider the transformed random variable $U = e^X$ having a thicker tail than distribution $F(x) = 1 - x^{-\alpha}$, $x > 1$, $\alpha > 0$ given in (2.1) of the reported Pareto variable X. Distribution of U has a slower decay of tail probability, viz., logarithmic decay $P(U > u) = (\log u)^{-\alpha}$, $u > e$; compared to polynomial decay in (2.1). The distribution has density of the form $f(u) = \alpha u^{-1}(\log u)^{-(\alpha+1)}$, $u > e$; which has a slower order decay than a Pareto density. As a result, the shift of restricted quantile under the condition of crossing a threshold is of higher magnitude than that for Pareto variable.

The shifted median under the restriction $U > u_0(\geq e)$ is at $u_0^{2^{1/\alpha}}$, the shifted p-th quantile is at $u_0^{(1-p)^{-1/\alpha}}$, $p \in (0,1)$.

The corresponding shifts for reported Pareto variable X mentioned (2.1) are $u_0 2^{1/\alpha}$ and $u_0(1-p)^{-1/\alpha}$, $p \in (0,1)$, $\alpha > 0$, these are of multiplicative order whereas that of the underlying unreported variable U are of power order.

Apart from the specific instance cited regarding alcohol consumption, the distribution of U may also be of interest to explain unaccounted gap between reported and unreported wealth. One may obtain Pareto type distributions from exponential distribution with random intensity following a beta prior, see Dasgupta (2011). A natural question arises whether it is possible to obtain from Bayesian consideration a distribution function with $P(U > u) = (\log u)^{-\alpha}$, $u > e$, having logarithmic decay; originating from Pareto distribution having polynomial decay. In the following we answer the question in affirmative.

Such a representation provides a Bayesian insight into the situation when a traditional model fails in favor of an alternative model.

Proposition 1. Let the random variable X be Pareto distributed with density function $g(x|a) = ax^{-(a+1)}, x > 1, a > 0$. For a fixed $a > 0$, let $U = e^X|a$. Suppose a has a prior gamma density

$f_{\beta,p}(a) = \frac{\beta^p}{\Gamma(p)} e^{-a\beta} a^{p-1}, \ \beta > 0, \ p > 0$.

Then the marginal density of X has similar decay as that of U, i.e., a monotonically decreasing density with decay lower than Pareto density,

$f(x) = p\beta^p x^{-1} (\log x + \beta)^{-(p+1)}, \ x > 1$.

Proof. Follows from integrating the joint density $g(x|a)f_{\beta,p}(a)$ with respect to a.

The marginal density of X remains bounded at $x = 1$ for every fixed $\beta > 0$. However, this blows up at the rate $p\beta^{-1}$ as $\beta \downarrow 0$. Height of relative histogram near left end point 1 may provide an estimate of p/β, mode of the distribution.

Proposition 1 has following implication. A typical heavy tailed distribution of a phenomenon may follow a Pareto model. However, aggregate of different groups having random Pareto indices following e.g., a gamma density may result in a heavy tailed distribution that may be more realistic in some situations.

3. Characterization theorems for discrete random variables

Consider a random variable X with support either $\mathbf{N_0}$, the set of nonnegative integers; or set of positive integers $\mathbf{N_1} = \mathbf{N_0} - \{\mathbf{0}\}$. Let the cumulative distribution function of X be denoted by $F(x) = P(X \le x)$, it is enough to define F at integer values. For $p \in (0, 1)$ the p-th quantile of F is defined as $F^{-1}(p) = \{\inf x : F(x) \ge p\}$.

The following two theorems are the counterparts of Theorem 1-2 stated for discrete random variables.

Theorem 3. For a random variable X with support $\mathbf{N_1}$ and distribution function $F(x) = P(X \le x)$, let the p-th quantile of the distribution under the restriction $X \ge x_0(\in \mathbf{N_1})$ be cx_0; where $c \in \mathbf{N_1}$ is the unrestricted p-th quantile of X. The above property holds for all p of the form $p = p_i = \sum_{j=1}^i P(X = j)$, $i = 1, 2, 3, \cdots$ iff $F(x) = 1 - x^{-\alpha}$ for some $\alpha > 0$, where $x \in \mathbf{N_1}$.

Theorem 4. For a random variable X with support $\mathbf{N_0}$ and distribution function F, let the p-th quantile of the distribution under the restriction $X \ge x_0(\in \mathbf{N_0})$ be $c + x_0$, where $c \in \mathbf{N_0}$ is the unrestricted p-th quantile of X. The above property holds for $p = p_1 = \sum_{j=0}^1 P(X = j)$, iff F is a geometric distribution function on $\mathbf{N_0}$.

Proof. Theorems 3-4 follow similar lines as that of Theorems 1-2. One way implications of the Theorems are easy to see. Consider the 'only if' part.

In the case of Theorem 3, steps similar to (2.2)-(2.4) hold. The variable X has support $\mathbf{N_1}$. This set is same as the set $\{c, c^2, \cdots, c^m, \cdots\}$, where $c = c(p)$ is the p-th quantile of X, and p of the form $p = p_i = \sum_{j=1}^{i} P(X = j)$, $i = 1, 2, 3, \cdots$. The p-th quantile is then an integer, as the jumps of F ocurr at integer points. For example when $F(x) = 1 - x^{-\alpha}$, the p-th quantile $c = c(p) = (1 - p)^{-1/\alpha}$ is obtained as the solution i of the equation $p = p_i = \sum_{j=1}^{i} P(X = j) = 1 - i^{-\alpha}$.

Over the set $\mathbf{N_1}$, characterization for $g(x) = \log \overline{F}(x) = -\alpha \log x$ is seen to hold in a similar fashion like in Theorem 1.

The proof for 'only if' part of Theorem 4 is similar to Theorem 2 along the above lines. However, note that in this case the set $\mathbf{N_0} - \{\mathbf{0}\}$ is also spanned by $\{c, 2c, \cdots, mc, \cdots\}$, where $c = c(p)$ is the p-th quantile of X, $p = p_1 = \sum_{j=0}^{1} P(X = j)$, i.e., $c = c(p) = c(p_1) = 1$. Thus the characterization for $g(x) = \log \overline{F}(x)$ holds with the solution $g(x) = -\lambda x$ over $\mathbf{N_0} - \{\mathbf{0}\}$, on the condition of restricted quantile for $p = p_1$ only. Since the total probability is 1, the probability mass at origin is taken care of and Theorem 4 holds.

Modeling with above two discrete distributions depends on the tail behavior of observed frequency distributions. Distribution $F(x) = 1 - x^{-\alpha}$ for some $\alpha > 0$, where $x \in \mathbf{N_1}$ may be termed as Discrete Pareto distribution. This may be an appropriate model for grouped Pareto variable, grouped over class intervals of equal length.

4. Bivariate exponential model and related distributions

Consider a bivariate exponential distribution, with exponential marginal. The relation $Y = X + Z$, $Z \geq 0$, is a special case of a more general model

$$Y = aX + Z, \ a > 0, \ Z \geq 0, \qquad \qquad ...(4.1)$$

where Z is independent of X.

This distribution has application in lifestyle data to explain the number of future physical relationships for an individual, given the past in a social environment where tie from past is loose, see Dasgupta (2011).

The restriction that the marginal distributions of X and Y are exponential with respective intensities λ_x and λ_y requires that the distribution of Z is of the form $a\rho + (1-a\rho)(1-e^{-\lambda_y z})$, $\rho = \lambda_y/\lambda_x = \mu_x/\mu_y$, $z \geq 0$. This is distribution of a random variable that is a product of two independent random variables - a Bernoulli random variable with mean $(1 - a\rho)$ and an exponential random variable with parameter λ_y. See Iyer, Manjunath, and Manibasakan (2002). The Bernoulli variable takes the value zero with probability $a\rho$, thus $Z = 0$, with a positive probability $a\rho$. Under this exponential model, the correlation between the two random variables are $r_{x,y} = a\rho$.

Now, the restriction $X > x_0 (\geq 0)$, shifts the quantiles of X by the same magnitude from unrestricted quantiles, and this imposes a restriction on the exponential random variable Y as $Y > ax_0$, hence the restricted quantiles of the latter random variable is shifted by ax_0, from the corresponding unrestricted quantiles.

The converse is also true, we have the following Proposition.

Proposition 2. Consider the model $Y = aX + Z$, $a > 0$, $Z \geq 0$, where Z is independent of X. Under the restriction $X > x_0 (\geq 0)$, let the quantiles of X be shifted by x_0. Let the resultant restriction $Y > ax_0$, shifts the quantiles of Y by ax_0. Then both the variables X and Y are exponential, and the distribution of Z is of the form $a\rho + (1 - a\rho)(1 - e^{-\lambda_y z})$, $\rho = \lambda_y / \lambda_x = \mu_x / \mu_y$, $z \geq 0$.

Proof. From the characterization of exponential distribution via conditional quantiles, it follows that both X and Y have exponential marginal. The result then follows from the assumed relation $Y = aX + Z$, $a > 0$, $Z \geq 0$, where Z is independent of X.

Next we investigate a bivariate Pareto model in terms of conditional quantiles. It is possible to obtain Pareto type distributions from exponential distribution with random intensity following a beta prior. The following result is proved in Dasgupta (2011).

Theorem A. Let the random variable X be exponentially distributed with density function $g(x|\theta) = (-\log \theta)\theta^x$, $x > 0, 0 < \theta < 1$, where θ has a prior beta density $f_{\alpha,\beta}(\theta) = \frac{\Gamma(\alpha+\beta)}{\Gamma(\alpha)\Gamma(\beta)}\theta^{\alpha-1}(1-\theta)^{\beta-1}$, $\alpha > 0$, $\beta > 0$.
Then the marginal distribution of X is approximately Pareto with monotonically decreasing density having polynomial decay
$f(x) = O_e((x+\alpha)^{-(\beta+1)})$, $x > 0$.

It may not be out of place to mention that in view of Theorem A along with Proposition 1, starting from exponential density it is possible to obtain a monotonically decreasing density with decay lower than Pareto density viz., $f(x) = O_e(x^{-1}(\log x + \beta)^{-(p+1)})$, $\beta > 0, p > 0$, $x > 1$; via a two stage prior of beta density and gamma density, as mentioned in Theorem A and Proposition 1. This step wise reduction provides a Bayesian insight when a candidate exponential model is replaced, from data viewpoint, by a heavy tailed distribution. Possible fluctuation of parameters over heterogeneous groups/items in a population, governed by beta and gamma distributions may explain such phenomena.
Consider the model (4.1). The intensities of Y and aX are λ_y and λ_x/a respectively. As in Dasgupta (2011) associate a beta prior $f_{\alpha,\beta}(\theta)$ of Theorem A, on $\theta = e^{-\lambda_y}$. In the r.h.s. of (4.1), this induces a prior on $e^{-\lambda_x} = e^{-\lambda_y/\rho}$, where $\rho = \lambda_y/\lambda_x$ is considered to be a constant. Integrating both sides of (4.1) with respect to the prior probability on θ, we then have the relationship, see Dasgupta (2011);

$$Y^* = aX^* + Z^*, \, a > 0, \, Z^* \geq 0 \qquad \qquad ...(4.2)$$

where the transformed variables X^*, Y^* have polynomially decaying densities as given in Theorem A.
For a Pareto variable the conditional quantile of the variable crossing a threshold is a constant multiple of the shift. Thus the conditional quantile of X^* under the restriction $X^* > x_0^*$ is approximately a constant multiple of unrestricted quantile, and this restriction on X^* imposes the restriction $Y^* > ax_0^*$ on Y^*, which is approximately a Pareto variable having polynomially decaying density. Thus the conditional quantiles of Y^* is also approximately a constant multiple of unrestricted quantiles. The variable Z^* is the product of two independent random variables - a Bernoulli random variable with mean $(1 - a\rho)$ and an approximately Pareto

random variable with same parameter as that of Y^*. Unlike the earlier case Z^* may not be independent of X^*, as conditional independence and marginal independence are not related in general. The parameter $\beta(> 0)$ quantifies the dispersed nature of the transformed variables obtained from original exponential distribution. Smaller the value of β, more dispersed is the transformed variable with heavy tail caused by diversity of individual intensities under consideration.

Such bivariate models are useful when value of one random variable is necessarily bounded below by the other, e.g., maximum diameter vs. minimum diameter of an approximate oval object in industrial production, number of relationships / physical encounters of an individual up to two successive time points from a common start, see e.g., Dasgupta (2011).
Observed frequency distributions may reveal sharp fall like exponential or, these may have relatively thick tails with approximate polynomial decay suggesting Pareto model. One may study simultaneous behavior of the conditional quantiles of two variables under the restriction of crossing thresholds, to search for an appropriate bivariate exponential or Pareto model.

In univariate case, there are situations when one is interested in studying the large values of the random variable with distribution having a thick tail, i.e., the behavior of the variable near the thick tail is of interest. In some cases Pareto model may provide a reasonable fit when the value of the variable exceeds some high threshold value. The Pareto fit is equivalent to constant multiplicative factor of restricted and unrestricted quantiles, former may then be computed in terms of the latter, thus providing magnitude of restricted quantiles indicating how large the variable can be near the tail.

5. Some examples

It is well known that Pareto distribution may explain the uneven distribution of wealth and income. Therefore the above mentioned property of constant multiplicative factor of restricted and unrestricted quantiles holds in such situations. We may examine the validity of such assumption in other situations. The above property regarding constant multiplicative factor of quantiles may hold for variable beyond a large threshold value. In such a situation Pareto distribution is appropriate above that threshold value.
The characterization provides the magnitude of shift in quantiles due to shift of threshold. In real life situations one may check the stability of shifted quantiles observed over several repetitions. Such stability of conditional quantiles of multiplicative form cx_0 in empirical distributions, crossing a threshold x_0 may indicate a Pareto model. The same may be said about exponential model by examining the stability of conditional quantiles of the form $c + x_0$ in empirical distributions, crossing a threshold.
In the following we check for Pareto model fit near the tail via R^2 of regression.

Example 1. *High tide water level at Arabian Sea*
High tide at sea causes tidal bore, a high tidal wave experienced in a narrow river or estuary that may cause substantial damage to lives and properties of inhabitants in nearby localities. Very high water levels are of concern.

The following data in feet, relates to high tidal range at Arabian Sea, west coast of India near Alang ship cycling yards. Each of these 170 observations was taken as maximum of two observations at different tide times, viz. at early hours and evening/night hours in a day. Thus, the observations represent the maximum height of sea water level in a 24 hour cycle. The data is spread over first six months in a year. Observation for a day is not taken into consideration, if any one of the two tide readings is missing in that day. The recorded observations are as follows.

36.39,36.65,36.46,35.70,34.39,32.45,30.32,30.22,30.62,31.44,32.32,33.01,33.40,
33.60,33.60,33.47,33.21,32.75,31.99,30.91,29.43,28.09,28.22,28.94,30.42,32.19,
33.96,35.50,36.59,37.08,36.95,36.10,34.49,32.16,30.88,30.09,30.06,30.58,31.27,
31.90,32.39,32.75,32.95,32.91,32.55,31.86,30.81,30.25,29.80,29.20,29.01,29.89,
31.60,33.54,35.21,36.39,36.82,36.46,35.31,34.36,32.95,31.14,29.47,28.71,28.94,
29.66,30.48,31.21,31.73,32.03,31.99,32.32,32.49,32.39,31.96,31.21,30.16,29.40,
29.83,31.27,32.98,34.42,35.28,36.36,36.72,36.26,35.08,33.31,31.24,29.24,27.99,
27.86,28.35,29.07,29.99,31.37,32.52,33.31,33.80,33.93,33.77,33.21,32.26,31.11,
30.25,30.22,30.98,32.39,34.62,36.23,37.01,37.01,36.29,34.95,33.31,31.44,29.60,
28.19,27.47,27.47,29.14,30.78,32.22,33.40,34.26,34.78,34.95,34.75,34.23,33.37,
32.35,31.37,30.68,31.44,33.50,35.18,36.23,36.59,36.36,35.60,34.52,33.24,31.86,
30.45,29.04,27.79,26.81,29.47,31.04,32.55,33.80,33.80,35.41,35.70,35.70,35.37,
34.68,33.60,32.19,30.55,32.22,33.60,34.65,35.28,35.41,35.21,34.72,34.06,33.31,
32.39

Figure 1 of $\log x$ vs. $-\log(1 - F(x))$ suggest that Pareto model may be appropriate beyond a large threshold value rather than the whole data set. In Figure 2 the same is plotted for $\log x > 3.55$ with 37 observations. The fit now seems better with squared value correlation as $R^2 = 0.8138$, and estimated value of $\alpha = 44.5849$.
With a further increase of the threshold value to $\log x > 3.58$ the Pareto fit (2.5) with $a = e^{3.58}$ to 20 observations seems more appropriate; providing the value of $\alpha = 102.6164$ from least square regression fit with a high value of $R^2 = 0.9207$.

Example 2. *Growth model for Elephant foot yam*
The following data relates to weights in kilogram of 100 yams from a growth experiment conducted in the year 2010 at Indian Statistical Institute, Giridih farm. We check the appropriateness of Pareto fit, especially beyond a threshold value.

4.50, 3.20, 2.60, 3.15, 2.05, 2.10, 2.65, 0.80, 1.70, 1.15, 2.90, 3.50, 4.35, 3.85, 3.60, 1.30, 2.20,
1.70, 3.70, 2.50, 3.40, 3.10, 4.45, 5.60, 4.15, 1.50, 1.90, 2.00, 3.10, 3.00, 3.10, 2.25, 2.65, 2.90,
3.60, 1.50, 1.20, 0.70, 2.80, 2.70, 3.75, 2.05, 1.60, 1.50, 3.60, 2.20, 1.40, 1.20, 0.00, 2.40, 2.50,
1.45, 1.05, 0.70, 0.00, 2.25, 2.00, 2.45, 1.55, 0.90, 0.75, 2.65, 2.25, 1.20, 2.25, 2.00, 3.80, 3.00,
3.00, 2.35, 1.05, 0.80, 3.80, 2.30, 3.80, 1.60, 0.00, 3.60, 1.60, 4.00, 3.00, 1.95, 2.00, 3.65, 3.60,
1.40, 1.40, 1.30, 3.90, 3.60, 5.50, 2.90, 2.60, 1.70, 2.80, 1.90, 1.70, 1.80, 1.10, 2.80.

In Figure 4 with 97 nonzero yam data we plot $\log x$ vs. $-\log(1 - F(x))$ and observe that Pareto model for Yam yield may be appropriate beyond a large threshold value, much like the earlier data on sea tide. Figure 5 plots the same for $\log x > 1$ with 38 observations. The fit of Pareto model (2.5) with $a = e$ now seems better as $R^2 = 0.9460$, estimated value of $\alpha = 5.7038$.

If the threshold value is increased slightly further to $\log x > 1.2$ as shown in Figure 6, we have 23 observations and a further increase in $R^2 = 0.9637$, providing a value of $\alpha = 7.0045$ for the model (2.5) with $a = e^{1.2}$.

One may compare the Pareto indices α_1, α_2 over two production scenarios, the smaller value of α signifies a better production; for in such case the (right) tail of the corresponding distribution is thicker compared to that with higher value. The ratio α_1/α_2 may serve as an index of production performance of situation 2 with respect to situation 1.

Example 3. *Peak gust wind velocities (PGU) in Florida, USA*
The following 156 observations relates to Peak gust wind velocities (PGU) for coastal city Florida, USA in miles per hour (mph) over 12 months recordings for several years during 1930-96. When peak gust wind velocities are not available, 5-second winds velocity preceding PGU are given. Wind types may be combined to reflect the highest reported wind velocity, see http://www.ncdc.noaa.gov/oa/mpp/wind1996.pdf for details.

41, 49, 41, 43, 61, 38, 41, 68, 68, 44, 85, 47, 52, 58, 77, 49, 69, 67, 67, 68, 48, 56, 47, 43, 52, 58, 48, 52, 62, 97, 69, 69, 74, 74, 68, 49, 40, 39, 46, 39, 40, 46, 45, 44, 92, 45, 31, 35, 55, 62, 66, 67, 56, 58, 69, 61, 55, 47, 46, 45, 58, 52, 75, 63, 52, 51, 51, 56, 58, 67, 69, 48, 45, 61, 59, 55, 46, 58, 56, 115, 62, 47, 49, 46, 48, 51, 62, 53, 68, 62, 74, 62, 56, 40, 41, 43, 54, 60, 59, 63, 60, 69, 64, 78, 79, 49, 69, 53, 35, 35, 35, 35, 32, 32, 35, 35, 53, 35, 35, 34, 44, 51, 53, 48, 41, 76, 67, 64, 83, 58, 68, 36, 44, 46, 58, 49, 51, 61, 60, 48, 45, 53, 60, 37, 46, 40, 43, 38, 39, 53, 32, 41, 52, 45, 46, 48.

In Figure 7 with 156 PGU data we plot $\log x$ vs. $-\log(1 - F(x))$ and observe that Pareto model for wind gust may fit well beyond a large threshold value, much like the earlier data on sea tide and yam-yield. Figure 8 plots the same for $\log x > 4.2$ with 31 observations. The fit of Pareto model (2.5) with $a = e^{4.2}$ seems reasonable as $R^2 = 0.9819$, estimated value of $\alpha = 8.0530$.

In Figure 9 we see that Pareto model to peak wind gust fits better for $\log x > 4.3$ with 13 observations and a high value of $R^2 = 0.9860$, with estimated value of $\alpha = 8.0128$ when $a = e^{4.3}$. The values of α seem to stabilize around 8, indicating stability of the model towards higher values of wind gust.

Taking $\alpha = 8$, the median wind gust exceeding the value $a = e^{4.3} = 73.70$ is $73.70 \times e^{\frac{1}{8} \log 2} = 73.70 \times 1.090508 = 80.37$ mph.

In a similar manner the median wind gust exceeding 115, the largest observation recorded in above PGU data, is $115 \times 1.090508 = 125.41$ mph.

This provides an idea about the magnitude of the variable *beyond* the reported records.

Example 4. *Worldwide earthquake data.*
The following data relates to earthquake measurements during 30 September - 1 October 2011 on Richter scale recorded worldwide,
see http://earthquake.usgs.gov/earthquakes/catalogs/eqs7day-M1.txt for details.
The webpage is continuously updated, and the following segment of data was collected some-

times on 1 October 2011.

1.7,4.8,2.2,2.8,1.8,2.3,1.2,2.1,1.4,3.0,4.8,3.6,1.6,1.3,2.4,3.8,1.3,1.7,4.7,2.8,5.4,5.2,3.1,
1.3,1.6,1.7,2.9,1.2,2.0,1.2,2.9,1.8,1.8,2.2,1.7,1.4,1.8,1.2,1.5,1.8,1.6,1.7,2.5,2.0,1.6,2.1,
1.0,1.3,3.3,1.7,4.4,2.9,1.6,2.5,2.8,2.6,1.3,1.9,2.0,1.6,1.1,2.5,1.3,1.1,1.2,1.2,1.1,1.4,2.4,
4.7,1.4,2.0,1.6,2.2,1.9,1.6,4.8,1.2,1.2,1.6,1.4,1.7,2.5,1.0,1.4,1.3,1.6,1.1,1.6,5.1,2.0,2.4,
4.5,2.5,2.8,1.7,1.1,3.2,1.4,1.5,2.4,1.2,4.8,1.4,2.0,4.8,1.3,1.8,4.6,1.0,1.8,1.3,2.5,1.1,1.9,
4.2,3.2,1.1,1.7,1.3,1.2,1.2,1.4,1.8,1.1,1.2,2.3,4.4,1.9,2.5,1.0,1.3,1.1,5.0,2.2,5.0,1.1,1.7,
1.3,2.9,1.2,1.0,1.9,1.8,2.1,1.8,1.6,1.6,2.7,1.1,1.5,1.8,1.1,1.4,4.6,1.6,2.0,2.5,1.4,1.8,1.1,
1.4,2.2.

In Figure 10 with 163 earthquake data we plot $\log x$ vs. $-\log(1 - F(x))$ and observe that Pareto model for earth-tremor may fit well as a mixture of two Pareto distributions. Figure 11 plots the same quantities for $\log x > 1.5$ with 15 observations. The fit of Pareto model (2.5) with $a = e^{1.5}$ seems reasonable as $R^2 = 0.9668$, estimated value of $\alpha = 18.0267$.
Pareto model to earth-tremor fit for $\log x < 1.5$ with 148 observations, is shown in Figure 12, $R^2 = 0.9378$, with estimated value of $\alpha = 3.0470$ when $a = 1.0$. The Pareto fit in lower range of the earthquake data is also satisfactory.
There seems to be a change in the parameter of distribution for tremor exceeding 4.2 in Richter scale. The physical interpretation of this is worth investigating.

References

Cebrian A, Denuit M, Lambert P (2003). "Generalized Pareto Fit to the Society of Actuaries Large Claims Database." *North American Actuarial Journal*, **7**, 18–36.

Dasgupta R (2011). "Discrete distributions with application to lifestyle data." *International Conference on Productivity, Quality, Reliability, Optimization and Modeling Proceedings, Allied Publishers, New Delhi*, **1**, 502–520.

de Oliveira MMF, Ebecken NFF, de Oliveira JLF, Gilleland E (2011). "Generalized extreme wind speed distributions in South America over the Atlantic Ocean region." *Theory of Applied Climatology*, **104**, 377–385.

Iyer S, Manjunath D, Manibasakan R (2002). "Bivariate exponential distributions using linear structures." *Sankhyā A*, pp. 156–166.

Jenkinson A (1955). "The Frequency Distribution of the Annual Maximum (or Minimum) of Meteorological Elements." *Quarterly Journal of the Royal Meteorological Society*, **81**, 158–171.

Klass O, Biham O, Levy M, Malcai O, Soloman S (2006). "The Forbes 400 and the Pareto wealth distribution." *Economics Letters*, **90**, 290–295.

Krishnaji N (1970). "Characterization of the Pareto Distribution Through a Model of Under-reported Incomes." *Econometrica*, **38**, 251–255.

Morrow-Tlucak M, Emhart C, Sokol R, Martier S, Ager J (1989). "Underreporting of Alcohol Use in Pregnancy: Relationship to Alcohol Problem History." *Alcoholism: Clinical and Experimental Research*, **13**, 399–401. Doi: 10.1111/j.1530-0277.1989.tb00343.x.

Van Montfort M, Witter J (1985). "Testing Exponentiality Against Generalized Pareto Distribution." *Journal of Hydrology*, **78**, 305–315.

Figure 1. Pareto fit for Tide data

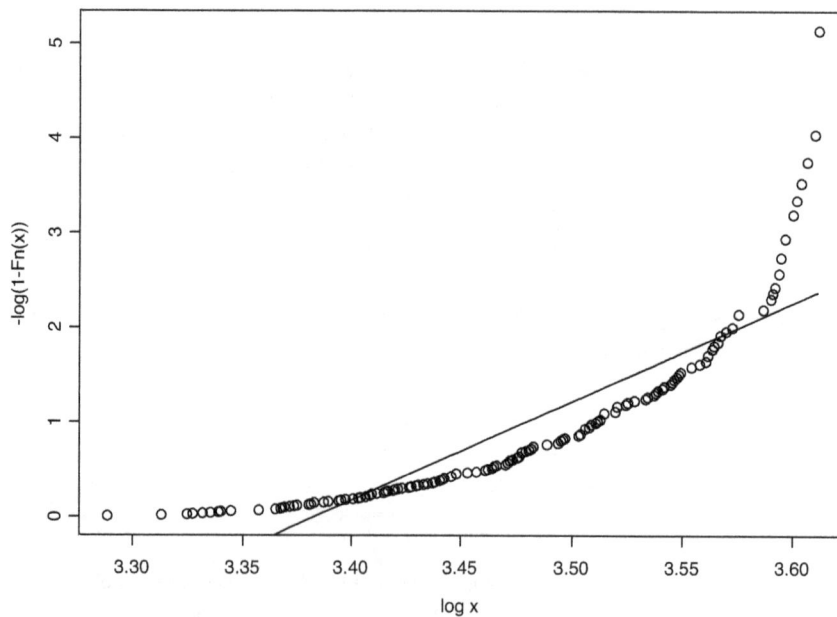

Figure 2. Pareto fit for Tide data: log X > 3.55

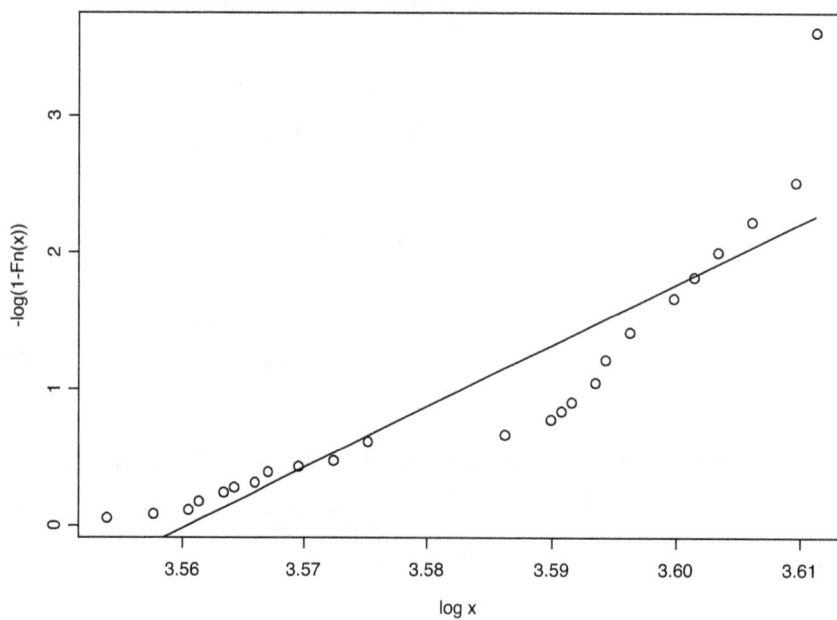

Figure 3. Pareto fit for Tide data: log X > 3.58

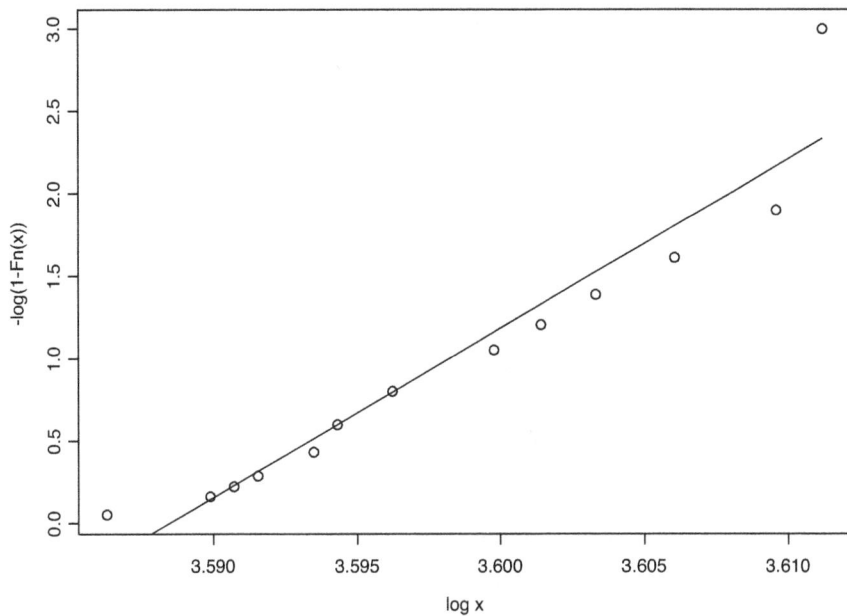

Figure 4. Pareto fit for Yam data

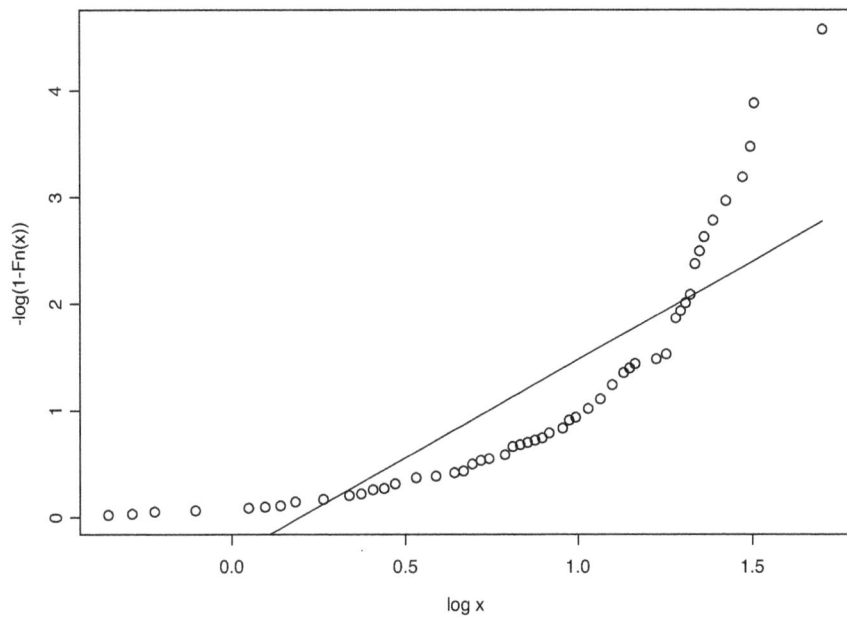

Figure 5. Pareto fit for Yam data: log x > 1

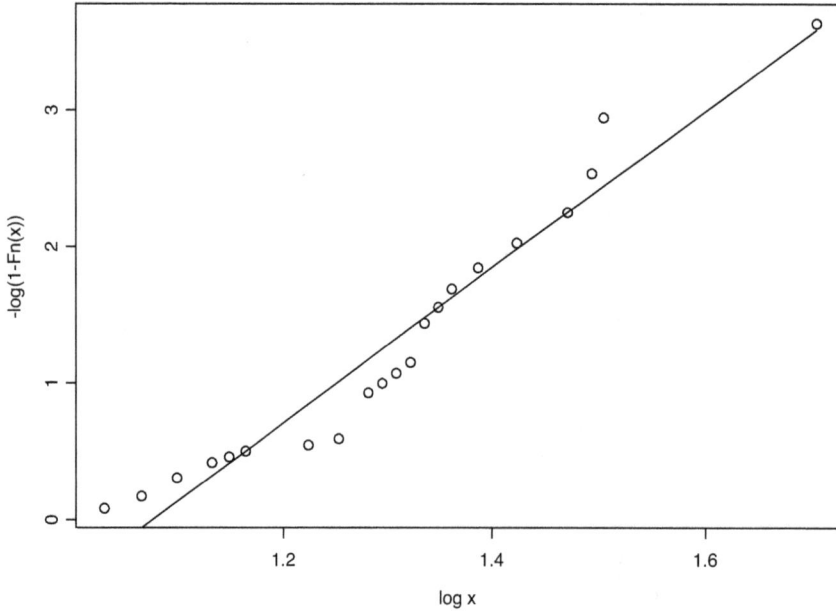

Figure 6. Pareto fit for Yam data: log x > 1.2

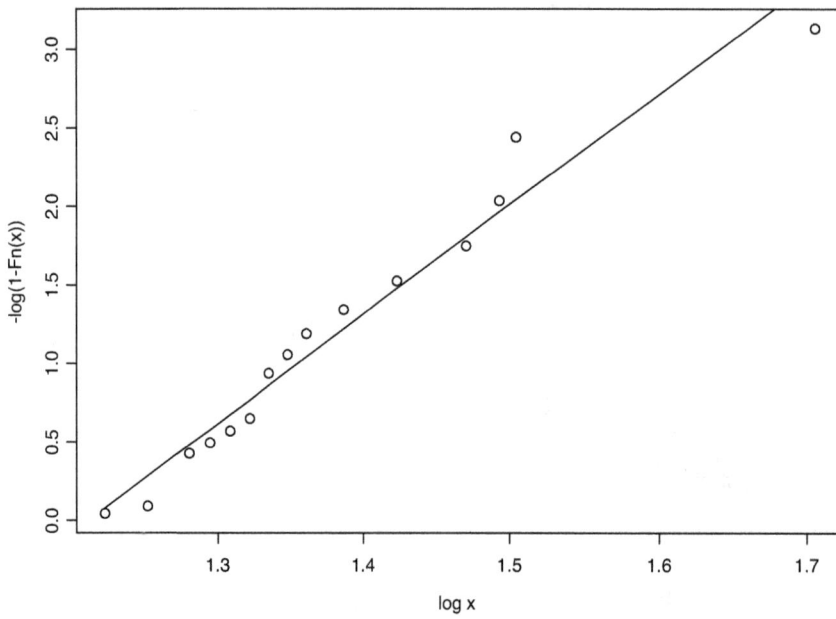

Figure 7. Pareto fit for peak wind gust data

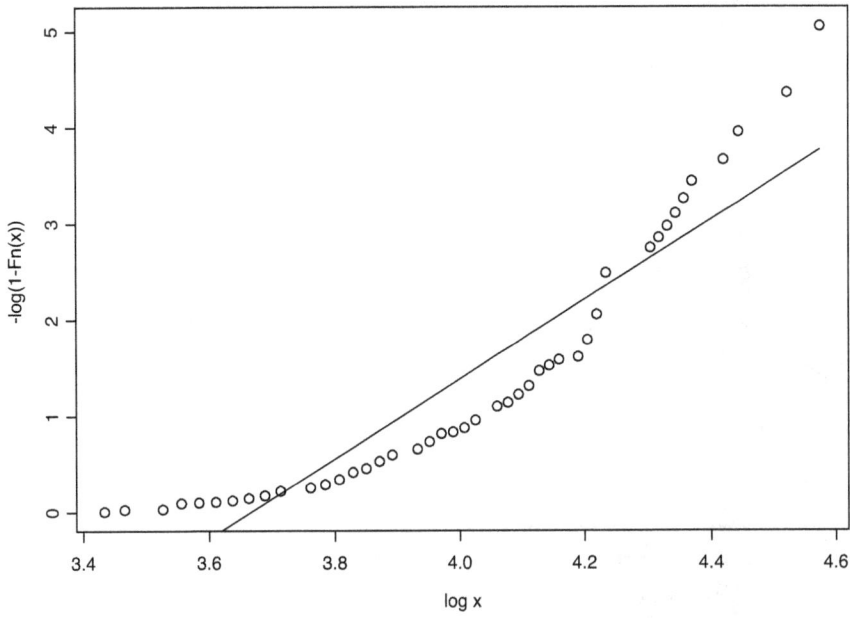

Figure 8. Pareto fit for peak wind gust data: log x > 4.2

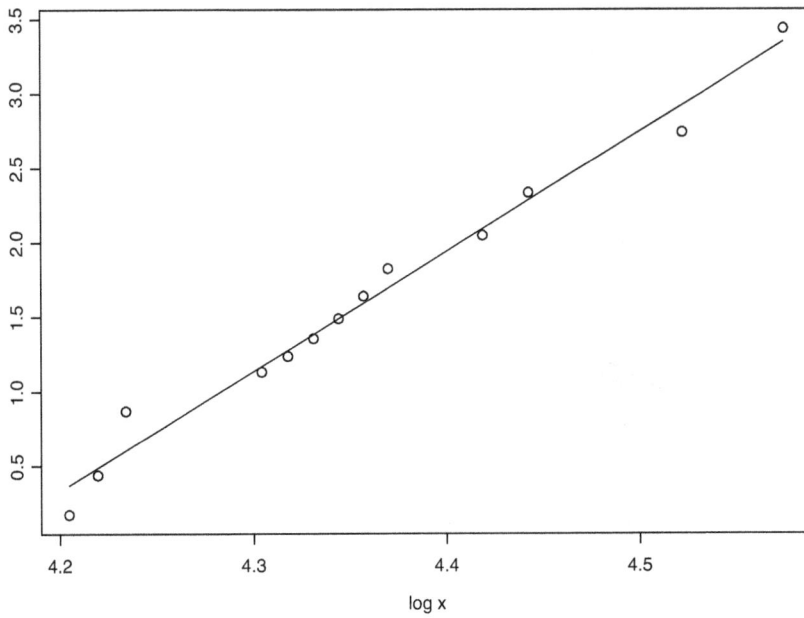

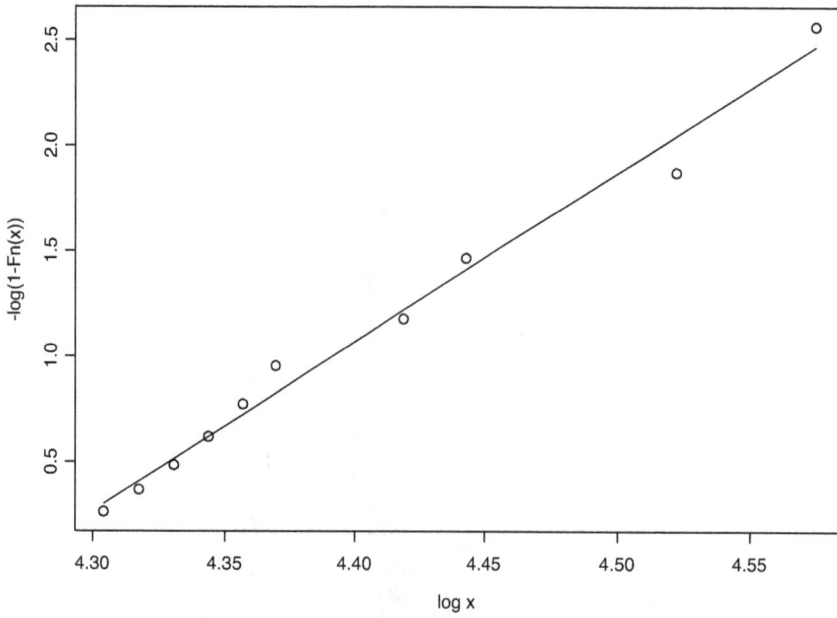

Figure 9. Pareto fit for peak wind gust data: log x > 4.3

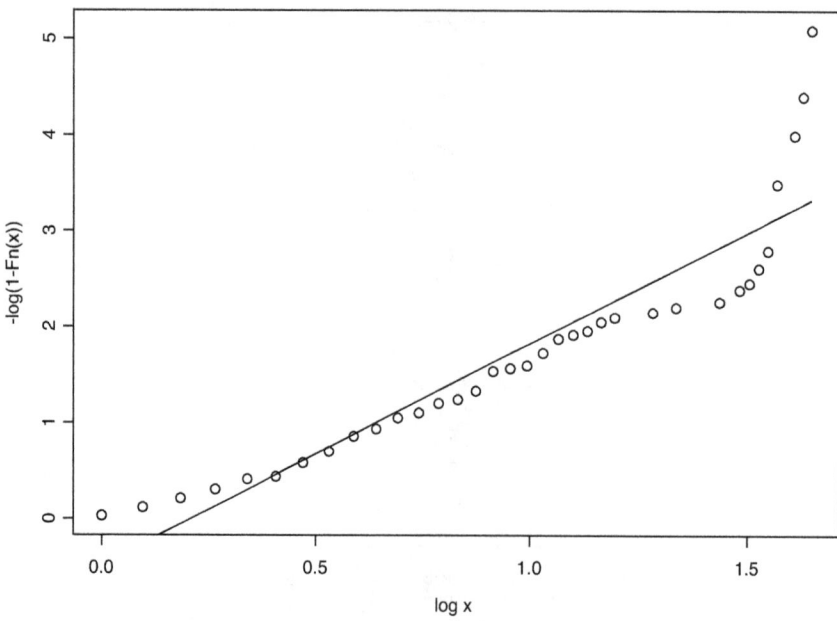

Figure 10. Pareto fit for earthquake data

Figure 11. Pareto fit for earthquake data: log x > 1.5

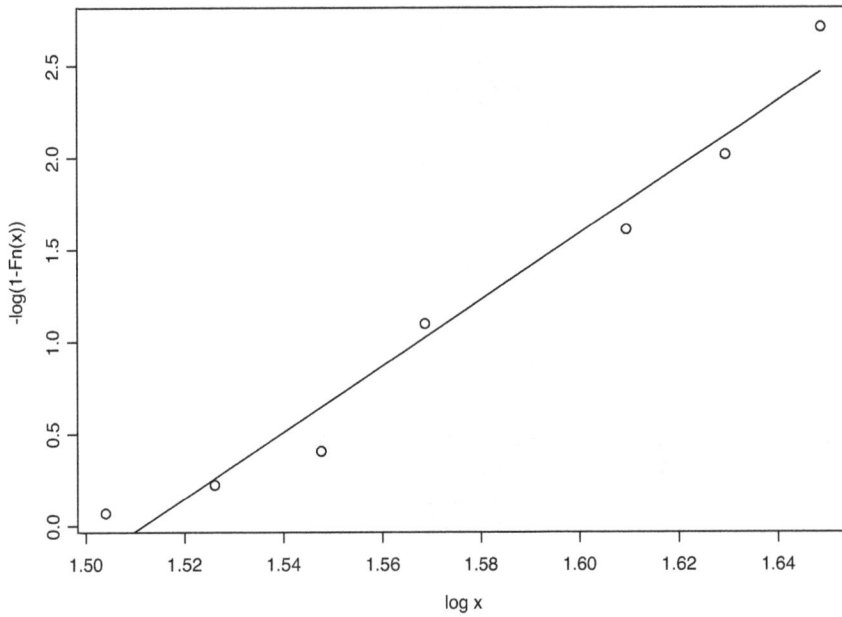

Figure 12. Pareto fit for earthquake data: log x < 1.5

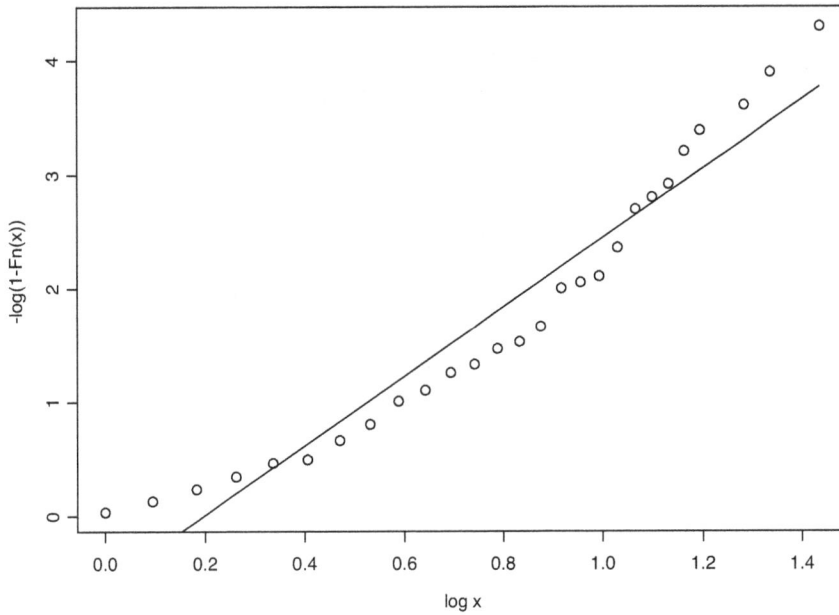

Affiliation:

Ratan Dasgupta
Indian Statistical Institute
Theoretical Statistics and Mathematics Unit
Calcutta 700 108, INDIA
E-mail: ratandasgupta@gmail.com
rdgupta@isical.ac.in

Permissions

List of Contributors

K.K. Jose
Professor, Department of Statistics, St.Thomas College, Pala, Mahatma Gandhi University, Kottayam, Kerala-686574 India

Bindu Abraham
Assistant Professor, Department of Statistics, Baselios Poulose II Catholicose College, Piravom, Mahatma Gandhi University, Kottayam, Kerala-686664 India

Craig Jackson
Department of Mathematics and Computer Science Ohio Wesleyan University Delaware, OH 43015

Sriharsha Masabathula
Department of Economics Ohio Wesleyan University Delaware, OH 43015

Lorena Vicini
Department of Statistics, UFSM Avenida Roraima, 1000, CCNE Santa Maria, RS, Brazil, 97105-900

Luiz Koodi Hotta
Department of Statistics, UNICAMP Rua Sérgio Buarque de Holanda, 651 Campinas, SP, Brazil 13083-859

Jorge Alberto Achcar
Department of Social Medicine Medical School of Ribeirão Preto Ribeirão Preto, SP, Brazil, 14049-900

A. Ian McLeod and Hyukjun Gweon
Department of Statistical and Actuarial Sciences The University of Western Ontario London, Ontario N6A 5B7 Canada

M. Saez
Research Group on Statistics, Applied Economics and Health (GRECS) and CIBER of Epidemiology and Public Health (CIBERESP). University of Girona, Campus of Montilivi, E-17071 Girona, Spain

M.A. Barceló
Research Group on Statistics, Applied Economics and Health (GRECS), and CIBER of Epidemiology and Public Health (CIBERESP). University of Girona, Spain

A. Tobias
Institute of Environmental Assessment and Water Research (IDAEA), Spanish Council for Scientic Research (CSIC), Barcelona, Spain, and Research Group on Statistics, Applied Economics and Health (GRECS), University of Girona, Spain

D. Varga
Research Group on Statistics, Applied Economics and Health (GRECS), and Geographic Information Technologies and Environmental Research Group University of Girona, Spain

R. Ocaña-Riola
Andalusian School of Public Health (EASP), Granada, Spain

P. Juan and J. Mateu
Statistics and Operations Research, Department of Mathematics, University Jaume I, Castell ón, Spain

Amira El-Ayouti and Hala Abou-Ali
Department of Statistics, Faculty of Economics and Political Science, Cairo University Cairo, Egypt

Jorge Mateu
Statistics and Operations Research, Department of Mathematics, University Jaume I, E-12071 Castellón, Spain

Jan R. Magnus
Department of Econometrics and Operations Research Vrije Universiteit Amsterdam De Boelelaan 1105 1081 HV Amsterdam, The Netherlands

Bertrand Melenberg
Department of Econometrics and Operations Research Tilburg University PO Box 90153 5000 LE Tilburg, The Netherlands

Chris Muris
Department of Economics Simon Fraser University 8888 University Drive Burnaby, BC V5A 1S6, Canada

Martin Wild
ETH Zürich Institute for Atmospheric and Climate Science CHN L 16.2 Universitätstrasse 16 8092 Zürich, Switzerland

Maria Franco-Villoria
School of Mathematics and Statistics

Marian Scott
University of Glasgow

Trevor Hoey
School of Geographical and Earth Sciences, University of Glasgow

Denis Fischbacher-Smith
Business School University of Glasgow

Benjamin Sexto M and Humberto Vaquera H.
Department of Statistics, Colegio de Postgraduados Campus Montecillo,Texcoco, Mexico 56230

Barry C. Arnold
Department of Statistics, University of California, Riverside Riverside, CA 92521

Alex Kostinski
Department of Physics Michigan Technological University Houghton, MI 49931

Amalia Anderson
Department of Physics Hendrix College Conway, AR 72032

HorováI., Koláček J. and LajdováD.
Department of Mathematics and Statistics Masaryk University Brno

María Pilar Frías
Department of Statistics and Operations Research, University of Jaén, Spain

María Dolores Ruiz-Medina
Department of Statistics and Operations Research, University of Granada, Spain

Ratan Dasgupta
Indian Statistical Institute Theoretical Statistics and Mathematics Unit Calcutta 700 108, INDIA

Index

www.ingramcontent.com/pod-product-compliance
Lightning Source LLC
Chambersburg PA
CBHW082028190326
41458CB00010B/3302